POCKET GUIDE TO

BUTTERFLIES

First published in 2015 by Bloomsbury Publishing Plc,
50 Bedford Square, London WC1B 3DP

www.bloomsbury.com

ISBN (print) 978-1-4729-1592-4
ISBN (ePub) 978-1-4729-1594-8
ISBN (ePDF) 978-1-4729-1593-1

A CIP catalogue record for this book is available from the British Library
Library of Congress Cataloging-in-Publication Data has been applied for

Publisher: Nigel Redman
Project editor: Jane Lawes
Design by Rod Teasdale

Printed in China

This book is produced using paper that is made from wood grown in managed sustainable forests. It is natural, renewable and recyclable. The logging and manufacturing processes conform to the environmental regulation of the country of origin.

10 9 8 7 6 5 4 3 2 1

POCKET GUIDE TO
BUTTERFLIES

Bob Gibbons

Phototographs by Bob Gibbons and Richard Revels

B L O O M S B U R Y
LONDON • NEW DELHI • NEW YORK • SYDNEY

CONTENTS

INTRODUCTION

WHAT IS A BUTTERFLY?

Butterflies are part of a large order of insects, the Lepidoptera, which also includes the moths. The characteristics of the whole order are wings covered in scales (lepidoptera means 'scale wings'), which give them their colour and pattern, and, normally, a long sucking proboscis that is coiled up under the head when not in use. This is used for drinking nectar and other fluids, the primary source of food for the adults.

Moths and butterflies differ in a number of ways. Within northern Europe all butterflies are day flying only, whereas most moths are night flying, although there are a number of day-flying moths that do resemble butterflies. All our butterflies have distinctly clubbed antennae, and there is a sharply defined, short swelling at the tip of each antenna. Most moths do not have this feature, but some, such as the day-flying burnet moths, have antennae that taper gradually from a swollen tip. Most moths rest with their wings folded back along their bodies, whereas most butterflies fold their wings above the body, revealing only the undersides. Finally, on detailed examination moths can be seen to have their forewings and hindwings attached together by little hooks, each called a frenulum, whereas butterfly wings simply have a large area of overlap. This combination of characteristics, and particularly the antennae, serves to distinguish the two groups.

Superficially, butterflies can also be confused with the owl-flies, or ascalaphids, which are active, day-flying insects in warm, grassy sites throughout southern and central Europe (but not the UK). They differ from butterflies in having very long, clubbed antennae and large, clear patches on the wings, and by their active predatory lifestyles.

THE LIFE OF A BUTTERFLY

All butterflies pass through a series of stages in their lives, known as complete metamorphosis. The females of adult butterflies lay eggs on specific food plants or groups of food plants, which they select; occasionally the eggs are deliberately laid away from the food plant, but near it. The eggs hatch to produce larvae, or caterpillars, which feed on the food plant, passing through a number of stages until they are fully grown. The larva is the main feeding and growing phase of the life cycle. When it reaches a certain size, a larva ceases feeding and turns into a pupa, or chrysalis, in a process known as pupation. This is an immobile phase and pupae are hidden somewhere attached to the food plant, in soil or in some other suitable place. Within the pupa an extraordinary process takes place, during which the body is effectively gradually reassembled into the form of an adult butterfly. When ready, the fully formed adult breaks out of the pupa. At first it is relatively small and shrivelled looking, unable to fly and vulnerable to predation. Within a few hours, however, the wings are inflated to their full size, and the butterfly is able to fly as a sexually mature adult.

The main function of the adult part of the life cycle is breeding and dispersal. Adults feed regularly on nectar and other nutritious liquids, primarily to gain energy, not growth. When not feeding, basking or sheltering from inclement weather, the males spend most of their time defending territories, seeing off potential rivals and seeking out females with which to mate. Females behave rather differently, spending much of their time searching for suitable plants on which to lay their eggs. Males are frequently more conspicuous than females for this reason, although not in all species.

Butterflies vary widely in their mobility and powers of dispersal. Some, such as the Duke of Burgundy and Grizzled Skipper (see pp. 54 and 12), live in small, loose colonies, with relatively little movement outside of these. Others, such as the Brimstone (see p. 52), are highly mobile, moving widely in search of mates, nectar or larval food plants, although they are not migratory in the sense of having regular defined movements over long distances.

A number of species are regular migrants, most commonly migrating northwards from a southern base (typically in the Mediterranean area) whenever populations build up. The pattern of these migrations varies widely. Some species perform reverse migrations (much has been discovered about this in recent years – see, for example, p. 118), while others do not return south. In some cases the migrants reinforce existing resident populations, while others only occur as migrants. In more extreme cases American species, notably the Monarch *Danaus plexippus*, appear in north-west Europe when they are blown off course from their regular extraordinary north–south migrations within the American continent. Nowadays there are examples of human-assisted migration; for instance, the Geranium Bronze *Cacyreus marshalli*, from South Africa, has become established as a resident in parts of southern Europe due to its introduction there with pot plants, and the availability of suitable garden plants as larval food. It is recorded occasionally in Britain, but has not become established.

FINDING AND WATCHING BUTTERFLIES

Although butterflies occur almost everywhere, it can be surprisingly difficult to find specific species without local knowledge. In general, the best way to discover a range of interesting and attractive butterflies is by visiting high-quality, semi-natural habitats such as chalk grassland (especially in June and July), old woodland (especially spring and summer), heathland (especially summer) and gardens at almost any time. All of these are rich in insect life unless they are badly polluted.

Nature reserves frequently preserve the best examples of these habitats, and most are open to the public. Some examples of organisations that have reserves are given on p. 189. It is also worth searching the internet as there are many websites that give details of good sites for particular species.

When out in the field it is best to walk slowly and scan ahead, to try and see the butterflies before they see you. If you approach slowly they will often not see you as a threat and may stay put. It is always a good idea to carry binoculars, particularly for viewing tree-based species such as some hairstreaks and Purple Emperors (see p. 144). Consider close-focusing binoculars (such as the Pentax Papilio, which is small, light and extremely close focusing, specifically for butterfly watching); these will allow you to examine the features of butterflies in detail.

In general, butterflies are easier to see but harder to approach in warm, sunny weather, when they are very active, whereas in cool or damp weather they are hard to find but easier to approach as they are reluctant to fly. In settled weather it is worth going out early as butterflies begin to warm up, or staying out late as they settle for the night or, in some cases, become more visible as they come down from trees.

HOW TO USE THIS BOOK

In order to identify a butterfly you have seen, the best starting point is to look through the photographs in this guide and try to find something similar to what you have seen. If it does not quite fit, check nearby pages, and also any similar species mentioned at the ends of species accounts. Also look at the information given on size, normal habitats, distribution in the region and flight period to try to confirm an identification.

FLIGHT PERIOD gives an indication of when the adult insect is likely to be seen. In particularly warm or cool sites, or in unusual years, adults may be seen outside the given period. In this era of climate change it is becoming increasingly difficult to give accurate flight periods, even for a given location. In general, there is a tendency for species to emerge earlier, and to be more likely to have an additional generation within the season than was the case in the past.

HABITAT AND DISTRIBUTION gives the normal habitats of the species – for example woodland – but many species are quite mobile and turn up in other habitats. Some details of where a species occurs in Britain and Ireland, and adjacent continental Europe, are given, but they are usually simplifications; many species are spreading or declining, so this information should not be considered as a defining feature. As in the case of flight period above, climate change is tending to alter the distribution of species, with many species moving northwards.

SIMILAR SPECIES gives an indication either of species that might be confused with the main one, or of some closely related species.

BUTTERFLY CONSERVATION

Insects in general and butterflies in particular are declining alarmingly in most developed countries. The combination of habitat loss, habitat fragmentation and the widespread use of pesticides and other damaging chemicals has led to huge losses of numbers, and some species have become extinct, at least locally, in recent decades. Tragic in itself, this decline has also affected our populations of birds, bats and other animals that are so dependent on insects.

Butterfly conservation has become more widespread and more effective in the last few decades than it used to be, because we have learned much about butterflies and their requirements, allowing targeted management in protected areas. This has revolutionised the management for some species, so that their populations have risen even though the number of sites where they occur has declined. There are also now some succesful examples of the reintroduction of locally extinct species such as the Large Blue (see p. 78) to their former haunts.

Several organisations (see p. 189) are specifically concerned with insect conservation, and of course the general nature conservation organisations help by protecting good habitats. Join as many of these as you can, and get involved.

To achieve something for butterflies personally, it is worth managing your garden with butterflies (and other aspects of nature) in mind. In summary, this includes planting or encouraging the larval food plants of likely local species, offering ample nectar plants for the adults, providing both shelter and warmth, and avoiding using any damaging chemicals. For more details, see the bibliography (p. 188).

SKIPPERS, HESPERIIDAE

These are quite distinct, small butterflies that are almost moth-like in appearance, with flattened and clubbed antennae. Some, the 'golden skippers', perch with their forewings and hindwings at different angles. There are eight regular species in the UK, and about 45 in Europe as a whole.

Dingy Skipper
Erynnis tages

Wingspan 25–30mm. A small, brownish-grey butterfly that is quite patterned when fresh but soon becomes duller. The underwings are paler yellowish-brown than the upperwings, but are rarely seen as the butterfly normally settles with wings open flat, or sometimes curled around a stem. Males and females are very similar, although the male has inconspicuous scent scales. The tiny greenish eggs are laid on Common Bird's-foot Trefoil, Horseshoe Vetch and other related legumes.

FLIGHT PERIOD Mainly late April–mid-June, with a small second brood in late summer in favoured localities if warm.

Dingy Skipper second brood adult on chalk grassland.

A courting pair of Dingy Skippers, male below.

HABITAT AND DISTRIBUTION Occurs in small colonies in warm, sheltered, neutral to calcareous locations such as downland, old railway cuttings and woodland clearings. Still quite common, although declining, in southern Britain, with scattered colonies as far north as the Moray Firth. Locally frequent in Ireland, including as a distinct subspecies in the west. Widespread in northern mainland Europe as far north as southern Scandinavia.

SIMILAR SPECIES Most likely to be confused with day-flying moths such as the Burnet Companion *Euclidia glyphica*, which has orange underwings and antennae without any swelling, or Grizzled Skipper (see p. 12), which has a more distinct chequered pattern, and chequered fringes to the wings.

Grizzled Skipper

Pyrgus malvae

Wingspan 24–28mm. A small and pretty butterfly, boldly chequered dark brown and white above when fresh, becoming duller as it ages, with chequered wing fringes. The undersides are paler than the uppersides, with bold white spots; they are quite often seen because this species frequently settles with its wings closed in dull weather or when roosting. Males and females are virtually identical, although males are more conspicuous and active, constantly settling and basking, then darting off to meet a female or see off another male, whereas females fly close to the ground looking for host plants. The roughly spherical, pale green eggs are laid singly on Wild Strawberry, cinquefoils and other members of the rose family.

FLIGHT PERIOD Mainly late April–end June, with a small second brood in late summer in favoured localities in warm seasons.

Grizzled Skipper larva feeding.

Grizzled Skipper adult basking.

HABITAT AND DISTRIBUTION Found in similar places to the Dingy Skipper (see p. 10), although it is less tied to calcareous soils. Typical sites include woodland clearings, banks, sheltered downland and valleys, generally characterised by shelter, warmth and vegetation that is not too luxurious.

SIMILAR SPECIES In Britain it is unlikely to be confused with much else when newly emerged, although worn individuals may resemble the Dingy Skipper; in nearby mainland Europe it is most likely to be confused with the Large Grizzled Skipper *P. alveus*, a variable and uncertain species that is always larger, with a wingspan of up to 34mm, and smaller and fewer spots. It occurs in similar habitats and is widespread, although it is rare or absent from Denmark and Holland.

Chequered Skipper

Carterocephalus palaemon

Wingspan 28–32mm. A particularly attractive little skipper with an appealing combination of large, orange-yellow spots on a warm dark brown background. The sexes are similar, although females are generally larger than males, with paler spots. Males normally pass their time perched with wings open in a sheltered spot ready to intercept any passing female, rival or even other insects, while females flutter in grass, or nectar at Bugle, bluebells and thistles. The pale, flattened, globose eggs are laid singly on the host plant, Purple Moor-grass, or other grasses such as Wood False-brome.

FLIGHT PERIOD Late May–early July in most localities; rather later in montane sites, in a single brood.

Chequered Skipper feeding on Thrift, showing underside.

Chequered Skipper basking, West Scotland.

HABITAT AND DISTRIBUTION Species of slightly damp, sheltered grassy places, such as woodland margins and clearings, and coppiced woodland, often on – although by no means confined to – quite heavy acid soils. Occurs in much of central Europe, and is most common at mid-altitudes in mountain areas. In Britain it is extinct in its former sites in central and eastern England, but still has strong populations locally in western Scotland. Elsewhere, it is widespread in central and western Europe as far north as northern Scandinavia, although it is absent from large parts of the flatter lowlands.

Large Chequered Skipper on Cross-leaved Heath.

SIMILAR SPECIES Within Britain, if you are lucky enough to see it, this species is unmistakable. However, in north-west Europe there are two rather similar species. The Northern Chequered Skipper *C. silvicolus* differs in that males are largely yellowish-brown above with a few dark forewing spots and dark shaded areas on the hindwings; females are more like Chequered Skippers, but have distinctly larger areas of yellow. This is a strongly northern species, becoming steadily more frequent from northern Germany northwards in damp meadows and woodland clearings. The Large Chequered Skipper *Heteropterus morpheus* is quite distinctive by virtue of its uniformly dark brown upper surfaces with just a few white spots on the forewings (more in the female than in the male), and with underside hindwings boldly marked with large, black-edged, oval white spots on a yellowish background. It also has a rather distinctive and curious bouncing flight, low over the ground. The species is locally common in sheltered damp meadows and wet heaths, from western France to southern Scandinavia in our area, flying late June–early August.

Mallow Skipper

Carcharodus alceae

Wingspan 30–35mm. A pretty but rather inconspicuous little butterfly, with marbled pale orange-brown and dark brown upperwing surfaces that have a slightly pinkish tinge, and distinctly crenellated hindwing margins. There may occasionally be small white spots or streaks on the forewings. Females are slightly larger than males, although otherwise the sexes are similar. The species settles frequently with wings either spread or closed, and the male often lowers its wings while raising its abdomen.

FLIGHT PERIOD One of the earliest species, appearing from March onwards, with several broods through to late summer; further south it may be present all year.

HABITAT AND DISTRIBUTION Occurs in a wide variety of sheltered, warm, flowery habitats. Recorded in Britain in 1923 but not since then; it was considered to be an introduction, although it can be quite a mobile species in warm weather. In continental Europe it is common south of 50 degrees north, and is currently spreading northwards due to a warming climate and widespread cultivation of potential host plants such as hollyhocks and other mallows.

SIMILAR SPECIES Distinctive in our area; the crenellated wing margins and lack of white helps to distinguish it from the Dingy Skipper (see p. 10) and similar moths.

Mallow Skipper basking.

Small Skipper
Thymelicus sylvestris

Wingspan 30mm. This is one of the five 'golden skippers' in the area, notable for their golden-brown colour and habit of settling with their forewings raised at about 45 degrees. The uppersides of the wings are golden-brown with faint black veins and shaded black margins. The sexes are similar except that the males have a distinct black sex brand running diagonally across the forewing. The undersides are unpatterned orange-brown shaded with greyish-green. Typically, males are more active than females, with the usual skipper behaviour of constant rapid forays to investigate females and potential rivals. In the UK the food plant is normally Yorkshire Fog, but elsewhere Timothy and other grasses are used.

FLIGHT PERIOD Flies in a single generation, early June–late August in the UK, although it may emerge much earlier in sites further south.

Female Small Skipper on Red Clover.

Male Small Skipper on Knapweed.

HABITAT AND DISTRIBUTION Occurs in rough, grassy areas such as roadsides, old railway land, neglected fields and downland, wherever the food plant is common; may become abundant in planted leys. Widespread and locally common throughout England and Wales, although less common in the north; currently expanding its range northwards.

SIMILAR SPECIES In general appearance and behaviour, resembles the other golden skippers (Essex, Large, Silver-spotted and Lulworth), being most like the Essex Skipper (for distinctions, see p. 20).

Essex Skipper

Thymelicus lineola

Wingspan 30mm. This little butterfly is very similar in almost every way to the Small Skipper (see p. 18), and was not recognised as a separate species until 1889. However, there are some distinct, small differences between the two species, and with practice it is easy enough to identify in the field. The most consistent and clearest difference lies in the antennae, which are boldly black tipped, particularly underneath, whereas in Small Skippers they are brown. To see this it is best to get below a settled butterfly and look up at it. In addition, Essex Skippers tend to be duller, paler brown, with more diffuse black margins, and the male sex brand is shorter and more exactly parallel to the front wing margin. There is a life-cycle difference, too: the Essex Skipper overwinters as an egg, while the Small Skipper overwinters as a larva. Essex Skippers particularly favour grasses such as Cocksfoot and Creeping Soft-grass in the UK, and use a wide range of grasses elsewhere.

Male Essex Skipper showing underwings.

Female Essex Skipper feeding.

FLIGHT PERIOD Single generation, generally emerging a little later than the Small Skipper, in late June, although the two fly together until late August.

HABITAT AND DISTRIBUTION Occurs in very similar rough, grassy habitats to the Small Skipper, including coastal marshes, where it may be abundant. In Britain it is almost confined to south-east England, where it is locally common, although there are increasing records from further west, including Wales and Eire. It is widespread and common in southern Europe, extending as far north as southern Sweden and the Baltic states.

Essex Skipper showing undersides of black-tipped antennae.

SIMILAR SPECIES The other golden skippers, especially the Small Skipper (see above).

Lulworth Skipper

Thymelicus acteon

Wingspan 25mm (smallest males); 26–28mm (females). This tiny little butterfly is the smallest of the golden skippers. Although it is broadly similar in appearance to Small Skippers, it differs in being generally darker and less orange. Females have a distinctive circle of orange rays on their forewings (occasionally present faintly in males), and males have a dark scent brand across the wing. The undersides are plain pale golden-brown. In sunshine the species is very active, and in large colonies every suitable flower (such as Marjoram or Greater Knapweed) will be crowded with nectaring butterflies. The larval food plant is Tor Grass.

FLIGHT PERIOD Single generation, although quite variable in its timing. In warm early colonies it can be seen from May onwards in good years, although normally it emerges in June and flies until early September.

Female Lulworth Skipper, showing the circle of orange clearly.

Male Lulworth Skipper on Scabious.

HABITAT AND DISTRIBUTION Within the UK this is an extremely local butterfly, still virtually confined to the county where it was first discovered – the south coast of Dorset, around Lulworth. Its stronghold lies in a coastal strip between Portland and Swanage, and it occurs abundantly wherever the larval food plant is allowed to grow tall, especially on warm undercliffs. Elsewhere in Europe it is common further south, but absent from the north of Holland and central Germany.

SIMILAR SPECIES The other golden skippers, but reasonably easily separated by its small size, dark colour and the female's golden 'eye'.

Large Skipper

Ochlodes sylvanus

Wingspan 32–36mm. This is the largest of the golden skippers, with rather longer antennae than other species. Both sexes are more boldly marbled with brown and orange than other golden skippers. Females are particularly strongly marked with orange and brown, and are larger than males, while the slightly duller males have a thick blackish sex brand running diagonally across the forewing. Males are very active in sunny weather, selecting a sunny perch from which they conduct numerous forays in search of females, or guarding their territory against other males. Females, by contrast, spend their time basking, feeding or searching for egg-laying sites on coarse grasses, especially Cocksfoot.

FLIGHT PERIOD The earliest of the golden skippers, with a single generation that lasts late May–early September (further south in Europe there may be a small second generation).

Female Large Skipper basking.

Male Large Skipper at rest. Notice the typical position of skipper wings at different angles.

HABITAT AND DISTRIBUTION Common and widespread throughout England and Wales in sheltered, rough, grassy sites with plenty of the food plant. Woodland rides and clearings are especially favoured. Rare in southern Scotland only, and absent from Ireland. Elsewhere in Europe it is common and widespread as far north as southern Scandinavia.

SIMILAR SPECIES The other golden skippers, but can be reasonably easily separated by its large size, marbled brown and yellow upperwings, and long antennae. The Silver-spotted Skipper (see p. 26) has more white on the upper surfaces, and its wings are boldly white spotted underneath.

Silver-spotted Skipper

Hesperia comma

Wingspan 29mm (males), to 36mm (largest females). A very attractive but surprisingly inconspicuous skipper. Although broadly similar in shape and general size to the other golden skippers, it differs in having much more white on the wings. The upper surfaces are marbled orange and brown, rather like those of the Large Skipper, but there are white spots on the forewings, especially on females. Males have a marked sex brand across the forewing. The undersides are particularly distinctive, boldly marked with large, silvery-white, squarish spots on a greenish-brown background. The butterflies settle either with their wings half-closed, in typical golden skipper fashion, or completely closed, when the underside spots are especially visible. The larval food plant is Sheep's Fescue.

FLIGHT PERIOD Latest of the golden skippers to emerge, with a single generation from late July (or even early August in cooler years) to early September.

The distinctive shining white egg of Silver-spotted Skipper on fescue.

Female Silver-spotted Skipper feeding on thistle.

HABITAT AND DISTRIBUTION Rare and local in the UK; confined to warm, usually south-facing calcareous grassland with bare areas and plenty of the larval food plant. Occurs only in south-east England within the UK, from Dorset to Kent, and north to the Chilterns. There are signs of a population recovery as a result of better management and global warming, but the species is very poor at dispersing to new sites. Locally common elsewhere in Europe, northwards to southern Scandinavia.

SIMILAR SPECIES The other golden skippers, but it can be reasonably easily separated from them because it has more white on the upper surfaces and boldly white-spotted undersides, and by its later flight period.

Male Silver-spotted Skipper feeding on Stemless Thistle.

SWALLOWTAILS AND APOLLOS, PAPILIONIDAE

This is a large and mainly tropical family, represented in Britain by just one resident species and a few occasional visitors.

Swallowtail

Papilio machaon

Wingspan to 80mm (males); 90mm (females). This is one of our largest, most beautiful and most distinctive butterflies; it is also one of the rarest. The sexes are similar, although females are slightly larger than males. The upper surfaces of the wings have a complex pattern of creamy-white patches on a blue-black background, with a line of blue spots and one red spot along the rear of the hindwings. Each hindwing ends in a prolonged 'tail'. British Swallowtails are a distinct subspecies, *P. machaon* ssp. *britannicus*, which appears darker than the continental subspecies (due to more extensive black markings), and has a different

A full-grown British Swallowtail larva on its food-plant.

Adult British Swallowtail on Ragged Robin.

lifestyle, wholly confined to marshes containing the sole food plant, Milk Parsley. These butterflies are strong fliers, frequently gliding, but the British populations tend not to leave their core area, whereas continental populations are very mobile. They regularly visit flowers such as Milk Parsley, Ragged Robin, Red Campion and Buddleia for nectar, and may readily appear in gardens within their area.

FLIGHT PERIOD The British subspecies has one generation per year (with an occasional partial second brood in August in warm years), late May–early July. Continental populations have two broods, and are almost continuously on the wing in May–September.

HABITAT AND DISTRIBUTION Very rare and local in the UK; confined to the Norfolk Broads in East Anglia, where it is locally common in suitable marshes and fens containing the food plant.

SIMILAR SPECIES The continental subspecies, *P. machaon* ssp. *gorganus*, is very similar, although generally a deeper yellow, and with fewer black markings. It is widespread throughout most of Europe except Britain and the far north, with mobile populations that can rapidly spread and colonise new areas in good years. It uses a much wider range of umbellifer food plants, including Fennel, Wild Angelica and Wild Carrot, so is much less restricted in its habitat types. This subspecies quite frequently wanders across to Britain, mainly to the south-east, and has occasionally become established for a while, for example in Kent and Dorset. Any swallowtail seen away from the Norfolk Broads is likely to be ssp. *gorganus*.

A continental Swallowtail, nectaring.

Scarce Swallowtail

Iphiclides podalirius

Scarce Swallowtail feeding on Asphodel.

Wingspan to 75mm (males); 86mm (largest females). A large, distinctive and beautiful butterfly. The upperwing surfaces are boldly marked with different-length diagonal black stripes on a creamy-white background, once described as looking like 'a small flying piano'. The hindwings have long dark 'tails', each with a paler tip. It is a strong flier, regularly migrating northwards in warm years, and using Blackthorn, Hawthorn and even some fruit trees as its larval food plant.

FLIGHT PERIOD There are normally two generations, the first flying mid-May–late June, and another late July–September.

HABITAT AND DISTRIBUTION Extremely mobile and likely to turn up in a wide range of flowery, open habitats, including parks, gardens and woodland glades. Widespread and common in southern Europe, but an uncommon vagrant in Britain, and in mainland Europe north of Holland and Germany.

SIMILAR SPECIES Unlikely to be confused with anything else in north-west Europe.

WHITES AND YELLOWS, PIERIDAE

These are medium to large white or yellow butterflies, variably marked with black spots.

Wood White

Leptidea sinapis

Wingspan 40–44mm. The smallest and most delicate of the whites, with roughly oval, rounded wings. The undersides – the only side normally seen in the field, as they always settle with wings closed – are yellowish-white suffused with a wash of grey-green scales. There is some variation between the sexes and generations, as well as a certain amount of geographical variation. Males have a well-marked dark wing-tip, which is greyish-black in the spring generation but darker in the second generation; females also have dark-tipped wings, but they are always paler than those of males. Summer-generation individuals are often smaller than those of the spring generation. Wood Whites from far western Ireland, in the Burren, have greener undersides than those in England.

A Wood White egg on the food-plant.

A male Wood White in flight, showing parts of the upper wing surfaces.

Wood Whites have a noticeably weak, fluttering flight, which is often sufficient to identify them by, and males spend much of their lives, given good weather, fluttering around in search of females. When a female is found, they embark on a curiously mannered, head-to-head courtship, which may last for five minutes or more before mating takes place.

Male Wood White on Yellow Meadow Vetchling.

FLIGHT PERIOD Normally two generations, although the second is usually smaller than the first. Spring generation emerges late April or early May and flies until late June; it is followed by a second brood in July and August. Further south in Europe there may be three broods.

HABITAT AND DISTRIBUTION Usually associated with woodland, in clearings, along rides or on partly shaded banks, where vetches such as Meadow Vetchling and Bitter Vetch occur. Very locally distributed in central southern Britain, from North Devon to mid-Wales, and in a small area of western Eire around the Burren in Clare and Galway. Common throughout most of mainland Europe except the far north.

SIMILAR SPECIES In 1988 it was discovered in the Pyrenees that there were two morphologically almost identical species of Wood White, but with different genitalia and courtship displays. Subsequent work confirmed this and showed that

Female Wood White, spring generation, with orchid pollinia attached to its proboscis.

the two species coexist in many places in Europe, but more recent research has split the species further. In the British Isles, interestingly, the wood whites of England, Wales and the Burren are all the original species. The wood whites of the remainder of Ireland – where they are largely grassland butterflies – were at first althought to be the Pyrenean Real's Wood White *L. reali*, but they are now distinguished as the Cryptic Wood White *L. juvernica*. The two species – the Wood White and Cryptic Wood White – do not overlap in Ireland, and the discovery explains the anomalous behaviour of the eastern populations, which tend to be more a species of grassland and other open habitats. There are subtle morphological differences between the two species, but examination of the genitalia is the only sure way to tell them apart. These three species are not yet well studied enough to give a precise distribution throughout Europe.

Black-veined White

Aporia crataegi

Male Black-veined White basking.

Wingspan 70–75mm. A large and elegant white butterfly. The wings are a clear pearly white, boldly marked on both surfaces with clear, thin black lines, as if drawn by a fountain pen. The sexes are similar, except that females tend to be more translucent than males. They are strong fliers, capable of flying long distances, and there may be mass movements northwards in high population years, although not as frequently as there used to be. The males in particular are frequent mud puddlers, with large numbers visiting damp, mineral-rich soil to obtain nutrients needed for the reproductive process. The food plant is most commonly Blackthorn, but apples, hawthorns and other rose-family shrubs may also be used.

FLIGHT PERIOD One generation emerges (according to latitude) in late April or May, with individuals lasting until late July.

HABITAT AND DISTRIBUTION Widespread and common in much of Europe as far north as southern Scandinavia, in open or semi-shaded, flowery habitats including meadows, gardens and wood

Black-veined White feeding on Bugloss, showing the bold black lines on the underside.

margins. Formerly known in the UK, it became extinct in about 1925 and has not recolonised.

SIMILAR SPECIES The Green-veined White (see p. 44) is much smaller, with more diffuse veins on a greenish background; the Clouded Apollo *Parnassius mnemosyne* can look similar, but has large black spots on the forewings, more translucent wings and a fatter body. It is local in Scandinavia and mountains further south and east.

Female Clouded Apollo settled on Bugle.

Orange-tip

Anthocharis cardamines

Male Orange-tip with wings open on Ladies' Smock.

Male Orange-tip with wings closed on garden bluebell.

Wingspan 40–44mm. The male Orange-tip is perhaps the most beautiful of all early spring butterflies, with its bold combination of orange tips to white wings above, and marbled blackish-green on white below. Females are largely white above, with narrow black tips to the wings and a black spot on each forewing (also present in males, but less noticeable), and similar to males below. Males have a gentle but persistent flight; they fly constantly in search of females or nectar when the sun is shining, but are rapidly grounded during periods of cold or wet weather. Females fly little, except to feed and lay eggs on the food plants, especially Garlic Mustard, Cuckoo Flower and other brassicas, which they find by a combination of visual and chemical stimuli.

Female Orange-tip with wings open.

FLIGHT PERIOD Normally one generation, which emerges in April (occasionally even earlier) and lasts through to mid-June; in exceptional years there is a small second brood in August.

Egg of Orange-tip on Ladies Smock.

HABITAT AND DISTRIBUTION A mobile species, found most frequently in damp meadows, wood margins and clearings, and gardens, wherever the food plants are common. Common in suitable habitats almost throughout the UK and Ireland, although very local in the far north. Widespread and generally common in most of the rest of Europe, reaching 2,000m or more in mountain areas.

SIMILAR SPECIES There is very little that is similar to the male in our region. Females are somewhat similar to the Bath White (see p. 46), but this is slightly larger, and has many more black spots on its upper and lower surfaces, and larger blotches of greenish colour on the underside.

Large White

Pieris brassicae

Wingspan 60mm (males); 68mm (females). This is our largest white butterfly, and is one of the two species (with the Small White) known generally as 'cabbage whites'. Both sexes have bright white wings with an extensive black tip to the forewing, but females differ in having two large black spots on the forewing (occasionally present as faint marks in males). The underwings are plain greyish-yellow. In general, the spring brood has paler black markings than later broods. These strong-flying butterflies are able to migrate over long distances, both northwards and southwards. The eggs are laid on members of the brassica family, especially cultivated ones, in groups of up to 100 on the undersides of leaves, and the resulting larvae can rapidly destroy their host plant.

FLIGHT PERIOD Normally two generations a year, with the first lasting from April (occasionally earlier) to June, followed by a more numerous second generation in July–September; in favourable conditions there may be a third autumn generation.

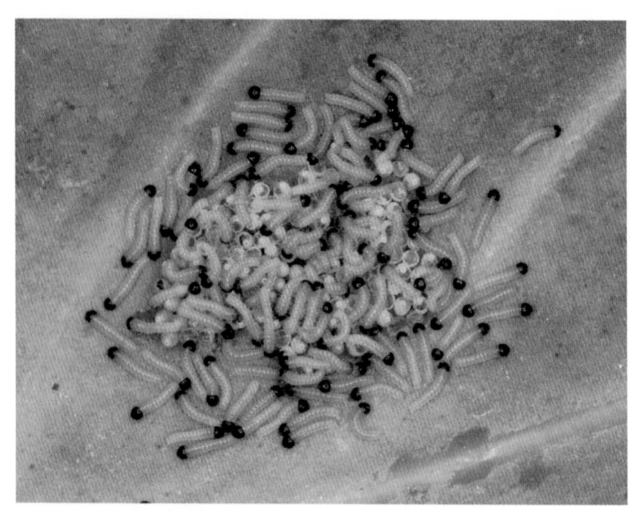

Large White caterpillars just hatching, on cabbage leaf.

Female Large White feeding on Knapweed.

HABITAT AND DISTRIBUTION Common almost everywhere. Does not really exist in colonies, but wanders widely in search of food plants using its sense of smell, so that almost any suitable area – particularly gardens and allotments – will be colonised in warm years. Resident throughout the UK, and supplemented by large annual inwards migrations from elsewhere in Europe.

SIMILAR SPECIES The Small White (see p. 42) is distinctly smaller, with less extensive black wing-tips, and males have a single black spot on the forewings; their eggs are laid singly, not in batches. Green-veined Whites and Orange-tips (see pp. 44 and 38) have quite differently patterned underwings.

Small White
Pieris rapae

Wingspan 40mm (small males), to 55mm (largest females). Although similar to the Large White, this species is much smaller. It is the other 'cabbage white' and can be a pest, although it is generally less damaging than the Large White as it frequently uses wild brassicas, and its eggs are laid singly. It is very similar in overall pattern to the Large White (for key differences, see p. 41). As in the Large White, the first generation each year tends to have fainter markings than succeeding generations. Although it is not a noticeably powerful flier, the Small White can cover large distances and mass migrations can occur in some years. It wanders constantly in search of food plants, and does not occur in discrete colonies. The summer brood is particularly attracted to blue flowers such as scabiouses.

FLIGHT PERIOD Normally two generations a year in the UK, April–June and July–September, with little gap between them; in favourable conditions there may be a third autumn generation. Further south there may be continuous generations from early spring to late autumn.

HABITAT AND DISTRIBUTION Common almost everywhere except in high mountains and on some islands. Like the Large White, this species wanders widely in search of food plants, so that almost any suitable area – particularly gardens and allotments – will be colonised in warm years. Resident throughout the UK, but supplemented by annual inwards migrations from elsewhere in Europe. Common in the rest of Europe with the exception of Iceland.

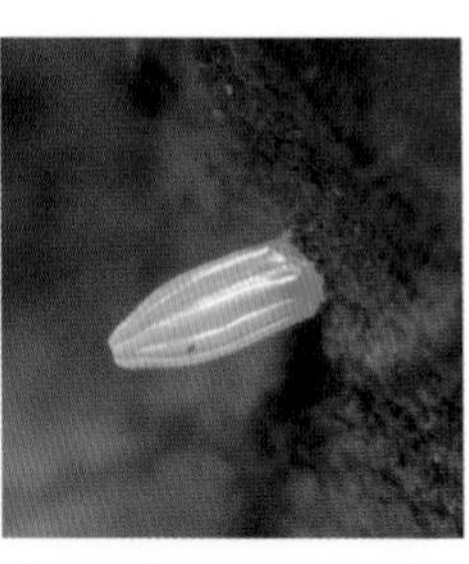

The single egg of a Small White, on cabbage.

SIMILAR SPECIES See under Large White for differences from that species. Green-veined Whites and Orange-tip females (see pp. 44 and 38) are similar in size and general colour, but have quite different underwing markings.

Mating pair of Small Whites.

Female Small White feeding on Cat's Ear.

Green-veined White

Pieris napi

Wingspan 45–52mm. Although similar in size and general colour to the Small White (see p. 42), this is definitely not one of the 'cabbage whites'. It is a more delicate butterfly, attuned entirely to wild members of the brassica family, especially Lady's Smock, Garlic Mustard and Water-cress, almost always growing in damp conditions. Its most distinctive feature, in both sexes and generations, is the pattern of the underside hindwing, which has all the veins outlined with a broad grey-green line, giving rise to the name 'green-veined'. First-generation males can be both smaller and more weakly marked than those of the second generation, giving rise to occasional confusion with Wood Whites (see p. 32), whereas first-generation females tend to be more boldly marked and veined than second-generation females. Despite its rather weak flight this is quite a mobile butterfly, covering large distances in search of food plants and mates, although it is not migratory like some other whites.

FLIGHT PERIOD Normally two generations a year in the UK, early April–June and July–September, with an occasional (but apparently increasing) small third generation. The second generation is generally slightly larger than the first.

A cluster of Green-veined Whites feeding on horse dung.

A male Green-veined White, spring generation.

HABITAT AND DISTRIBUTION Locally common, occasionally abundant, throughout the UK, although more patchily distributed than the cabbage whites due to its reliance on damp habitats such as fens, marshes and damp woodland, and their plants. Common in the rest of Europe with the exception of Iceland and a few other islands.

SIMILAR SPECIES Seen from below it is distinctive; from above the veins are often more edged than in the Small White, and some forewing veins end in expanded dark triangles. The Wood White has more oval wings, is usually smaller, and has more extensive dark-tipped wings and no black spots on the wings.

Bath White

Pontia daplidice

Wingspan 50mm. This extremely mobile migratory species, centred around southern Europe, periodically migrates northwards into northern Europe, including Britain. It is roughly the same size as a Small White (see p. 42), and is distinguishable from other similar-sized species in this area by two key features – the extensive pattern of merging black spots on the upperwings, especially the forewings, and the attractive greenish marbling on the undersides. It looks immediately blacker above than any comparable white (almost more like a Marbled White, see p. 180), and greener below. It does not form colonies, but wanders widely and continuously in search of mates and food plants; these latter include various brassica-family plants such as Sea Radish, White Mustard, Hedge Mustard and the unrelated Wild Mignonette, all of which grow in open, sunny areas.

FLIGHT PERIOD In southern Europe it can be on the wing almost continuously, late March–October. Most likely to be seen in Britain in July and August, although since its peak in 1945 it has been a rare and sporadic visitor only.

Female Bath White on Scabious.

HABITAT AND DISTRIBUTION

Most likely to be found in open habitats with annuals, including the coast and wasteland. Common in Mediterranean Europe, but much more sporadic and infrequent in northern Europe, where it appears as far north as southern Sweden, in almost any open flowery habitat.

SIMILAR SPECIES In northern Europe only likely to be confused with female Orange-tips (see p. 38). These are marked with much less black above, and on the undersides the green colour is more mixed up with black and white. Elsewhere in Europe there are more similar species.

Pupa or chrysalis of Bath White.

Female Bath White feeding.

Clouded Yellow

Colias croceus

Wingspan 52–60mm. One of our most beautiful butterflies which, despite the warming climate, is still essentially an irregular visitor to Britain. From its base in the Mediterranean region (where it may occur all year in the warmest parts), it migrates northwards in variable numbers. Nowadays individuals are seen in Britain every year, but there are undoubtedly good years (such as 2000, 2006 and 2013) and bad years (such as 1999 and 2007), when numbers vary widely.

As the butterflies normally settle with wings closed, it is the undersides that are seen; these are bright orange-yellow (forewings) or greenish-yellow (hindwings), with a few black spots on the forewing, and a small white 'snowman' in a larger spot in the centre of the hindwing. The upperwings are deeper orange-yellow, extensively edged with black. There is also a common form of the female, var. *helice*, which is very pale yellow or greyish-white, with an orange spot on each hindwing. The food plants include fodder clovers, Lucerne and various wild legumes such as vetches and Bird's-foot Trefoil.

The pale *helice* form of a female Clouded Yellow.

A mating pair of Clouded Yellows.

FLIGHT PERIOD Most likely to arrive in Britain in May or June (occasionally earlier), and quickly establishes breeding populations that last through to October. Some individuals migrate back, but normally there is no overwintering population apart from tiny populations on the south coast.

HABITAT AND DISTRIBUTION This is a very mobile butterfly without fixed colonies. It is most likely to be seen in warm, flowery places such as chalk downland, or where there are concentrations of the food plants. It is one of the few species that do well in improved fields. Although most frequent in the south, it can turn up anywhere.

SIMILAR SPECIES Other clouded yellows (for differences, see p. 50), and Brimstone (see p. 52).

Pale Clouded Yellow

Colias hyale

Berger's Clouded Yellow

C. alfacariensis

Wingspan 52–60mm. These two clouded yellows are very uncommon migrant visitors to Britain and other parts of northern Europe. They are extremely similar to each other, and differ from the Clouded Yellow (see p. 48) in having lemon-yellow upperwings edged with black, and paler yellow, less greenish underwings. Attempting to separate the two species is extremely difficult, as the morphological differences are slight.

There are three rather variable characteristics to look for: the Pale Clouded Yellow has a more pointed apex to the forewing than Berger's. From close up you can see that the orange spot on the upper surface of the forewing is much brighter orange in Berger's than in Pale; and the grey dusting of scales near the body is more extensive in Pale. None of these traits is diagnostic, and often you have to just accept that it could be either species. There are some differences in behaviour. Berger's lays eggs only on Horseshoe Vetch (or occasionally other species) and is more sedentary, whereas Pale uses similar food plants to the Clouded Yellow and wanders widely. The larva is quite different – that of Berger's is pale green, with two yellow lines and black spots in each segment, whereas that of Pale has no spots, fewer lines and a darker green colour.

FLIGHT PERIOD Most likely to be seen July–August, and may be offspring of earlier arrivals. Pale is the more common species of the two.

HABITAT AND DISTRIBUTION Pale may be found almost anywhere in the south, while Berger's usually turns up on chalk downland.

SIMILAR SPECIES Clouded Yellow (see p. 48).

Berger's Clouded Yellow feeding on Knapweed.

The distinctive caterpillar of Berger's Clouded Yellow.

Brimstone

Gonepteryx rhamni

Wingspan 60–70mm. A beautiful, large and distinctive butterfly. Both sexes have distinctively shaped wings, resembling a leaf, with a marked point on each wing; they always rest with wings closed. Males are bright butter-yellow, while females are pale greenish-yellow, and both have a single orange-brown spot in the centre of each wing. This is a mobile and strong-flying species, but it is not typically migratory. It wanders widely in search of nectar and the food plants, Buckthorn and Alder Buckthorn, but populations are not reinforced by migrants from elsewhere. After emergence from hibernation in spring, the butterflies tend to stay in the habitats where the food plants occur. From late July onwards the new generation spreads out in search of warm nectaring places, and later hibernation sites.

FLIGHT PERIOD One of the few British butterflies to pass winters as an adult butterfly. Hibernated individuals emerge on warm days in spring (although they may appear during any warm day in winter), whereupon mating and egg laying take place. The new generation emerges in late July and flies until hibernation in autumn.

The large distinctive eggs of Brimstone on Buckthorn.

An overwintered male Brimstone.

HABITAT AND DISTRIBUTION In spring, most likely to be seen in woodland clearings and gardens, and along hedges, wherever the food plants occur, on both acid and calcareous soils. The new, late-summer generation is more wide ranging than the old one, turning up in any sheltered flowery sites. In the UK the species is strongly southern, virtually absent from Scotland and local in Ireland. It is widespread in northern continental Europe.

SIMILAR SPECIES The Clouded Yellow (see p. 48) may look similar in colour, but is smaller, and has a different wing shape and extensive black markings above. Pale females may resemble Large Whites (see p. 40), but the leaf-shaped wings and lack of any black spots help separate the Brimstone from this species.

METALMARKS, RIODINIDAE

This is a largely tropical family with just one European representative. Nowadays it is often considered part of the much larger Lycaenidae.

Duke of Burgundy

Hamearis lucina

Duke of Burgundy caterpillar on Cowslip leaf.

Wingspan 30mm. A pretty little butterfly rather like a fritillary in colour and pattern, but more like a blue in shape and size. The upper forewings of both sexes are dark brown chequered with orange, while the hindwings are duller brown, with a few pale markings. Both wings of both sexes have dark dots around the margins. Although the sexes are similar, females generally have a brighter, bolder orange chequer pattern, and males have reduced, useless forelegs. The underside hindwings have two bold bands of white dots. This is a very sedentary butterfly, rarely moving far, with poor colonisation abilities. It has declined dramatically in recent decades to less than 200 colonies in the UK, and this is related partly to its inability to colonise new sites. Its food plants are Cowslips and Primroses.

FLIGHT PERIOD In the UK normally flies in one generation, late April–early June, although in 2007 a small second generation was noted. Elsewhere in Europe second generations are more frequent.

HABITAT AND DISTRIBUTION In Britain now virtually restricted to sheltered downland sites and woodland edges, although once it was frequent in woodland clearings (and still is elsewhere in Europe). Most common in southern England on chalk or limestone, but also with a few strongholds in northern England. Common in central Europe, but rare in both the north and south.

Above: Male Duke of Burgundy basking.

Right: Female Duke of Burgundy egg-laying on Cowslip.

SIMILAR SPECIES Really too small to be mistaken for a fritillary, and the black spots all around the margin, or the white spots below, help distinguish it. See also Chequered Skipper (page 14). Also has a distinctive behavioural trait, in which the males congregate in a lek around a group of suitable bushes.

COPPERS, HAIRSTREAKS AND BLUES, LYCAENIDAE

These butterflies are part of a huge family worldwide. In northern Europe they are all small, active butterflies, often with a metallic sheen on the wings, and many live in association with ants during their pre-adult phases.

Small Copper

Lycaena phlaeas

Wingspan 28mm (smallest males); 35mm (largest females). A small, bright and active little butterfly. Males and females are very similar, except that females are larger; both have beautiful metallic coppery orange-brown forewings, with black dots and a black margin, while the hindwings are dark brown with a broad orange trailing edge. Underneath, the forewings are paler orange with black spots, and the hindwings are dull greyish-brown. There is also a lovely form with a line of blue dots on the hindwings, most common in the north. In sunny weather males establish a sunlit perch from which they make constant forays in search of females, or to intercept rivals. Females are less active, feeding, basking or searching for the larval food plants, Sorrel or Sheep's Sorrel.

FLIGHT PERIOD Flies in 2–3 broods, first appearing April–early June, then July–August, and frequently with a third brood running through until late October in warm years.

The eggs of Small Copper.

HABITAT AND DISTRIBUTION Species of warm, flowery, open areas where the food plants occur, such as grassy heaths, chalk downland, damp meadows and wasteground. Widespread throughout the UK except in mountains, although usually in small numbers. Widespread in almost all of continental Europe.

Third brood Small Copper basking in autumn sun.

Small Copper feeding on Yarrow, showing underside of wings.

SIMILAR SPECIES The female Large Copper (see p. 58) has a similar pattern, but is much larger and has silvery grey under hindwings.

Large Copper

Lycaena dispar

Wingspan 44–50mm. The Large Copper – especially the freshly emerged male – is a strikingly beautiful butterfly. It used to occur in the UK as a distinct and larger subspecies, ssp. *dispar*, which became extinct in 1864. A Dutch subspecies, ssp. *batavus*, is quite similar, and has been introduced to eastern England on several occasions, but no colonies have persisted and it is also extinct in the UK. The two subspecies depend, or depended, on Great Water Dock as their food plant. A third subspecies, ssp. *rutilus*, is widespread in central Europe, where it feeds on a wider variety of docks. Males are a beautiful bright coppery-orange above, with black-edged wings, while the slightly larger females have many large black dots on the forewings, and largely dark hindwings; both have silvery grey under the hindwings, edged with an orange stripe at the rear.

FLIGHT PERIOD Dutch subspecies flies in one brood, July–August; ssp. *rutilus* has two generations, June–September.

Large Copper, female *batavus* in flight.

Female Large Copper *batavus* feeding on Purple Loosestrife.

Male Large Copper *batavus* on Purple Loosestrife.

HABITAT AND DISTRIBUTION Dutch subspecies is very rare in fens; ssp. *rutilus* has many scattered sites from western France eastwards. It is absent from most of Scandinavia.

SIMILAR SPECIES Male Scarce Coppers *L. virgaureae* are similar to male Large Coppers, but a little smaller, with white-spotted, brownish underwings. They occur in similar habitats, from southern Scandinavia southwards.

Brown Hairstreak

Thecla betulae

Eggs of Brown Hairstreak at the node of a Blackthorn bush.

Wingspan 36–44mm. This is the largest and latest of the hairstreaks. From below (the side most often seen), males and females are similar, both patchily golden-brown with two wavy white lines on the hindwings, although females have more orange on the forewings. In thin sunshine they may bask with their wings open, revealing dark brown wings in the male, while females have a large orange flash on each forewing. Both sexes have 'tails' on the hindwings.

The butterflies are found at low densities in areas where the primary food plant, Blackthorn, occurs. They are known to use individual large master trees, especially Ash, as an aid to finding a mate – all the local population gathers there soon after emerging. Binoculars are a useful aid to finding them at this period. Males are particularly sedentary, but females may come down in search of nectar from thistles, Bramble flowers and Wild Parsnip, or of rotting fruits such as those of Wayfaring Trees.

FLIGHT PERIOD Flies in a single brood, late July–September, although in warm years and/or sites, individuals may linger later.

HABITAT AND DISTRIBUTION An insect of sheltered, intricate countryside, typically with abundant Blackthorn hedges, ample large trees and numerous flowers, although it by no means occurs in all apparently suitable sites. In Britain virtually confined to south-central England and West Wales, with some colonies in western Ireland. Elsewhere widespread in central and western Europe, as far north as southern Scandinavia.

SIMILAR SPECIES The rich orange underside with two white lines serves to distinguish it from other woodland hairstreaks.

Female Brown Hairstreak basking with wings open.

Above: Male Brown Hairstreak basking with wings open.

Right: Female Brown Hairstreak with wings closed.

Purple Hairstreak

Favonius quercus

Wingspan 35–40mm. An attractive although rarely seen little butterfly. The underwings of both sexes are similar, an attractive dove-grey with a single wavy white line, and a conspicuous orange-and-black spot close to the 'tail'. Males' wings are wholly purplish-blue, apart from at the margins, with the degree of colour varying according to the angle of view; females are a duller blackish-blue, apart from a large blotch of purple sheen on each forewing. Despite this beautiful colouration, the butterflies are easily missed because they spend much of their time in and around oak trees, which are also their larval food plant. Binoculars are a useful aid to seeing them. They feed largely on aphid honeydew on tree leaves, so have no need to visit flowers, although they may occasionally do so, especially if Alder Buckthorn or Blackberry are in flower. They also bask readily with open wings, although frequently too high up to be visible.

FLIGHT PERIOD Flies in a single brood, concentrated in July–August, although individuals may be seen earlier or later.

HABITAT AND DISTRIBUTION The larval food plants are almost exclusively native oaks, and although it can occur on isolated oaks, it is much more likely to be found in and around oak woodland. It is quite a common species and easily the most common hairstreak, although it is easily overlooked. Within Britain most frequent in England and Wales, petering out northwards and westwards. Widespread throughout Europe as far north as southern Scandinavia.

Underside of Purple Hairstreak feeding on Alder Buckthorn flowers.

SIMILAR SPECIES None.

Boldly coloured female Purple Hairstreak basking.

Male Purple Hairstreak basking.

Green Hairstreak

Callophrys rubi

Wingspan 30–34mm. A delightful little spring-flying butterfly. When you see pictures of a Green Hairstreak it looks rather conspicuous, but in the field it is easily overlooked – dull brownish in flight, and like a leaf when it settles. The undersides are similar in both sexes, largely bright iridescent green with a single interrupted white line on the hindwing (and occasionally on the forewing). The upper surfaces, which are rarely seen in the field, are dull brownish in both sexes, differing only in that males have a small, pale sex brand on each forewing. Males are highly territorial, adopting a sunny perch and making forays to see off rivals or check females. Females spend more time fluttering among potential food plants, which include a wide range of species, especially Gorse, Bilberry, rock-roses, Bird's-foot Trefoil and Blackberry.

FLIGHT PERIOD Flies in a single brood, late April–early July. In warmer parts of Europe there may be a small second brood in late summer.

HABITAT AND DISTRIBUTION Due to its wide variety of food plants, occurs in various habitats on different soil types. Most such sites are warm, sheltered, flowery and often sloping, with abundant shrubby vegetation; they include railway banks, heathland edges and woodland clearings. Widespread almost throughout Britain and Ireland where suitable habitats exist, although rarely common. Occurs in most other parts of Europe.

SIMILAR SPECIES None in the area.

Larva of Green Hairstreak feeding.

Green Hairstreak on Hawthorn.

Green Hairstreak perched on a Marsh Orchid.

White-letter Hairstreak

Strymonidia w-album

Wingspan 30–36mm. An inconspicuous little butterfly. The undersides are all that is seen in the wild, as it always settles with wings closed; they are dark greyish-brown marked with a jagged white line that usually includes a 'W' shape near the tail – this is the 'letter' of both the English and the scientific names. The margin of the hindwing has a continuous band of orange crescents, and there is a distinct prolonged 'tail'. The uppersides are dark brownish, although they are never seen in the wild except for a glimpse when it flies. The butterflies spend most of their lives in and around the food plants – various elm trees – feeding largely on aphid honeydew, but at times they may come down to feed on thistles, Privet, Blackberry blossom and other flowers.

White-letter Hairstreak feeding on umbellifer flowers.

FLIGHT PERIOD Flies in a single brood, peaking in July, although individuals may be found in late June and early September.

HABITAT AND DISTRIBUTION Sedentary, occurring in definite colonies that are wholly tied to the presence of suitable elm trees, especially more mature ones that flower. These may be in woodland margins, clearings or roadsides. An uncommon species that has declined due to the demise of many elms. Most frequent in south-central England, and absent from Scotland and Ireland.

SIMILAR SPECIES The Black Hairstreak (see p. 68) is similar, but generally paler, lacks the distinct white 'W' shape, and has a row of black spots along the inner margin of the orange band.

White-letter Hairstreak feeding on thistle.

Black Hairstreak

Satyrium pruni

Black Hairstreak pupa on Blackthorn leaf.

Wingspan 35-40mm. An inconspicuous and rather dull-coloured butterfly. Only the undersides are normally seen; they are brown with a golden sheen when fresh, marked with a thin, jagged white line, and edged with a row of black dots along a broad orange edge. The upper surfaces are dark brown edged with orange blotches on both wings (female), or just the hindwings (male). The butterflies fly very little, spending most of their time in bushes and trees, where they feed on honeydew, but they become more mobile on warm evenings, coming down to flowers such as Privet or Blackberry, or searching for Blackthorn plants. The sole larval food plant is Blackthorn and its close relatives.

FLIGHT PERIOD Flies in a single brood, mid-June–mid-July, with a slight tendency for emergence to become earlier.

HABITAT AND DISTRIBUTION A very sedentary butterfly occurring in definite colonies with little movement into new areas. Its food plants are common and widespread, and its limited distribution must be partly accounted for by its poor dispersal powers. Most likely to be found around mature Blackthorn on woodland edges, rides and sheltered meadows. Within Britain it is very local, confined to a belt of woodland in the East Midlands. Elsewhere in northern Europe it is widespread except in Scandinavia, but rarely common.

SIMILAR SPECIES White-letter Hairstreak (for key differences, see p. 67). The Ilex Hairstreak *S. ilicis* is similar and usually rather greyer, and in place of the broad orange band edged with black dots it has a few scattered orange and black lunules. It is common in warm, scrubby places with oaks, especially evergreens, from southern Sweden southwards.

Black Hairstreak female, perched.

Ilex Hairstreak feeding on Tufted Vetch.

Long-tailed Blue

Lampides boeticus

Wingspan 33–40mm. A medium-sized blue butterfly. The upper surfaces are greyish-blue with a variable amount of purple sheen; the undersides are pale greyish-brown with transverse streaks of white and a broad white band towards the margin. There are two black spots in orange on the hindwing, close to the long 'tail'. This is an extremely mobile and widespread species around many of the warmer parts of the world, and is known as a pest of Broad Beans in places, but it is unable to survive in cooler areas.

FLIGHT PERIOD In its southern European stronghold this species is on the wing for much of the year, and certainly in February–November. Further north, where it is resident, its numbers peak in late summer. In some years it migrates strongly northwards, reaching as far as southern Britain and northern Germany; individuals in these areas are most likely to be seen August–October.

HABITAT AND DISTRIBUTION Rare non-resident visitor to Britain, although records seem to be becoming more frequent, and 2013 was an exceptionally good year. Most likely to be seen in open, sunny, flowery places, especially where food plants such as peas, beans and shrubby legumes are to be found.

Long-tailed Blue feeding on Forget-me-not.

SIMILAR SPECIES Unlikely to be confused with anything in this area.

Long-tailed Blue perched with wings open.

Small Blue

Cupido minimus

Wingspan 18–28mm. This is our smallest native butterfly. The undersides of both sexes are similar – plain silvery-grey with a scattering of small black dots, but no orange. The upper surfaces vary slightly – those of males are plain charcoal-black with a slight dusting of blue scales, while females are more dark brown, and both have a clear white fringe. Small Blues are less active than many butterflies, often basking and only flying short distances, so they are easily overlooked, especially as they are both small and dark. They often settle with wings half-open, giving a useful glimpse of both their upper and lower surfaces. Their food plant – Kidney Vetch – is used for both laying and nectaring, so much of their lives are spent on and around it, although usually they go elsewhere to a shrub to roost.

FLIGHT PERIOD Main generation late May–early July, although in warmer sites there is often a small second generation in August.

Male Small Blue, basking.

Mating pair of Small Blues.

HABITAT AND DISTRIBUTION Warm, sheltered places on light soils where Kidney Vetch thrives, such as chalk quarries, downland and dunes. Widespread throughout Britain, but rare everywhere away from the far south. Locally common in much of northern Europe as far north as southern Sweden.

SIMILAR SPECIES Nothing very similar, although the undersides of the larger Holly Blue (see p. 74) have a similar pattern.

Holly Blue

Celastrina argiolus

Larva of Holly Blue, attended by ant.

Wingspan 28–36mm. A distinctive little blue butterfly that is quite different from other blues in both appearance and behaviour. The undersides are silvery-grey with no orange spots and a pattern of small black spots, rather like in a large Small Blue (see p. 72). The upperwings are an attractive bluish-violet, more or less plain in males, but boldly edged with black, especially on the forewings, in females. First-generation females are less black than second-generation ones; females are also a pale powder-blue in the first generation, and more purplish in the second generation. They are unusual, among blues in the UK, in that the food plants are both shrubs, and are different for each generation. The spring generation lays eggs on Holly buds, while the autumn generation does so on Ivy. Unlike other blues the butterflies are not colonial, wandering widely in search of mates and food plants, and any blue butterfly flying high among shrubs and small trees is likely to be this species.

FLIGHT PERIOD Usually the first blue to appear, late March–late May, then again in late summer.

HABITAT AND DISTRIBUTION This wide-ranging species turns up wherever the food plants occur, especially in parks and gardens, and around woodland margins. In Britain it is most common in the south and absent from Scotland. Although it is often common numbers vary considerably, partly in response to the population cycles of a parasitic wasp. Common in most of continental Europe, where it sometimes uses quite different food plants from those used in Britain, such as White Melilot.

SIMILAR SPECIES Underneath rather like a large Small Blue, although quite different above. The underside separates it from all other northern European blues.

Female Holly Blue on Holly.

Holly Blue, late summer brood, perched.

Short-tailed Blue

Cupido argiades

Wingspan 22–30mm. A pretty little butterfly not much larger than a Small Blue (see p. 72). The upper surfaces are bright violet-blue in males, and rather browner in females. The undersides resemble those of the Small Blue, except for two conspicuous orange spots near the hind margin, and a short, thin 'tail'.

FLIGHT PERIOD In its core area further south in Europe, flies in two generations, April–June and July–August. In Britain and northern Europe, where it only occurs as a migrant, most likely to be seen late July–September.

HABITAT AND DISTRIBUTION A very rare immigrant to southern England, also reaching scattered locations from northern Germany to Finland. Most likely to be seen in rough meadows, flowery woodland margins and glades, especially where the food plants such as Red Clover and Lucerne occur.

SIMILAR SPECIES Small Blue (see p. 72).

Short-tailed Blue drinking, showing the distinctive underside.

Green-underside Blue

Glaucopsyche alexis

Wingspan 35mm. A medium-sized blue butterfly most noticeable for its undersides, which are charcoal-grey, but with a strong suffusion of greenish-blue over much of the hindwing. The forewing has a bold, curved arc of black spots, with a smaller, paler echo of them on the hindwing. The upper surfaces are bright blue in males, and brownish in females with a bluish tinge. The wing fringes are plain white, and not chequered. The food plants are various herbaceous legumes.

FLIGHT PERIOD Single brood, April–early July.

HABITAT AND DISTRIBUTION Occurs in a wide range of open, flowery habitats such as meadows, both dry and damp, woodland clearings and margins. Widespread in much of Europe as far north as southern Sweden and Finland; absent from the UK.

SIMILAR SPECIES Mazarine Blue (see p. 86).

Green-underside Blue perched on Green-winged Orchid.

Large Blue

Maculinea arion

Wingspan 40–44mm. Although not particularly large, this lovely blue is just bigger than the other UK blue species. From being quite common in Britain in the 19th century, it declined to the point of extinction by 1979 due to a combination of habitat loss, changed management and possibly over-collecting. At just about the time when it was becoming extinct, its extraordinary association with a particular species of ant, *Myrmica sabuleti*, was being elucidated, and this work has subsequently informed a major successful reintroduction effort using insects from Sweden. This is the only blue in Britain to have substantial areas of large black spots on the upperside forewings, especially in the female. The underwings are greyish-brown with extensive areas of black, white-ringed spots on both wings. The main food plant is Wild Thyme.

FLIGHT PERIOD Single brood, early June–mid-July.

Female Large Blue basking.

Female Large Blue egg-laying on Thyme.

HABITAT AND DISTRIBUTION Restricted to dry, warm grassland sites on acid or calcareous soils where the main food plant and the specific ant both occur. Such sites are usually grazed. In Britain all populations come from reintroduced butterflies, but it is doing well in some areas within south-west England. Elsewhere in Europe it does best in hilly areas, as far north as southern Sweden and Estonia.

SIMILAR SPECIES The closely related Alcon Blue *M. alcon* has no (or very few) black spots on the upper surfaces, and rather smaller spots on the undersides. It lives in damp meadows, with Marsh Gentian as its main food plant.

Silver-studded Blue

Plebejus argus

Wingspan 30mm. An attractive little blue butterfly. Males have deep silvery-blue upperwings, with broad black borders edged with a clear white fringe; females are chocolate-brown above, sometimes tinged with blue, edged with orange spots, especially on the hindwings. Below, they are typical of most blues, with a ground colour of greyish-brown dotted with black spots and edged with a stripe of merging orange spots; within the outer black spots there are silvery blue-green pupils – the silver studs of the name – which are diagnostic to this species in Britain (although not elsewhere).

FLIGHT PERIOD Single brood, early July–end of August, although on some sites, especially on limestone, it may emerge earlier.

Male Silver-studded Blue, perched, showing the silver 'studs'.

Male Silver-studded Blue basking on Bell Heather.

HABITAT AND DISTRIBUTION The largest populations in Britain occur on southern lowland heaths, where the food plants are heather and Gorse, but it is also found on limestone, where it uses rock-roses and Bird's-foot Trefoil. A local and declining species now absent from northern England, Scotland and Ireland, although there is a strong population of a different form, *caernensis*, in North Wales. Common in most of continental Europe in a wide range of habitats.

SIMILAR SPECIES In Britain the dark border and silver studs are diagnostic. Elsewhere in northern Europe the Idas Blue *P. idas* is very similar, differing in having (on average) a narrower black border, smaller underside spots and no spine on the male front tibia. It is widespread in mainland Europe in similar habitats; absent from the UK.

Brown Argus

Aricia agestis

Wingspan 27–31mm. A pretty little butterfly despite its dull colours. The upper surfaces of the wings are rich chocolate-brown edged with bold orange spots, which are particularly conspicuous on the female; in males, the forewing spots fade towards the front. Below, the underwings are broadly similar to those of other blues, pale greyish-brown with black spots and edged with orange spots. In fact, on close examination this species differs from all the other UK blues in having no spots on the half of the forewing nearest to the body (see Common Blue, p. 88, for example; it has two spots in this area). Both sexes resemble females of other blues, such as the Common Blue, but they lack any hint of blue in their colour. The primary food plant is Common Rock-rose, although it may also use small cranesbills and storksbills.

Male Brown Argus basking, on rock-rose.

FLIGHT PERIOD Flies in two slightly overlapping generations, May–June and July–September. Further south in Europe there may be three generations.

HABITAT AND DISTRIBUTION Particularly associated with warm, sunny, dry calcareous sites, such as downland, where rock-roses do well, although also occurs on dunes and neutral grassland. Locally common in much of England; rarer and largely coastal in Wales, and absent from Scotland and Ireland. Common in much of northern mainland Europe, although virtually absent from Scandinavia.

SIMILAR SPECIES Most likely to be confused with female Common Blues (see above), and the Northern Brown Argus (see p. 84).

Female Brown Argus – note the more extensive forewing spots.

Northern Brown Argus

Aricia artaxerxes

Wingspan 27–31mm. A controversial little butterfly whose status has changed frequently from species to subspecies to variety of Brown Argus, but recent genetic work has clearly established its full species status, although its entire distribution is not yet fully elucidated, and there are probably a number of distinct races within the species. The butterfly resembles the Brown Argus (see p. 82) closely, and is usually most easily distinguished by a prominent white spot in the centre of each forewing (although this may be reduced or absent in some populations). In Britain, and generally in northern Europe, there is virtually no overlap in the distribution of this species with the Brown Argus, which helps with identification. It also has some different behavioural characteristics, such as laying eggs on the upper surface of the food-plant leaf (underside in the Brown Argus), and almost invariably only having one generation, from early June to August (although a few northerly Brown Argus populations may also only have one generation).

Male Northern Brown Argus, basking.

Northern Brown Argus, showing underside of wings.

FLIGHT PERIOD Flies in one generation, early June–August, starting rather later in northern colonies.

HABITAT AND DISTRIBUTION Locally common on calcareous soils where rock-roses occur, such as rough grassland, cliffs, old quarries and lower mountain slopes, from north-west England northwards. In continental Europe widespread in Scandinavia and on mountains further south.

SIMILAR SPECIES Brown Argus (see above).

Mazarine Blue

Cyaniris semiargus

Male Mazarine Blue basking.

Wingspan 32–38mm. A medium to large blue butterfly. From above it is not particularly distinctive. Males are an attractive deep blue with narrow dark margins – rather like large Silver-studded Blues (see p. 80) – while females are uniformly deep brown with a slight bluish flush. Underneath, the butterflies are more distinctive, with a rather yellowish grey-brown base colour, sometimes described as cinnamon, and an arc of black, white-edged spots, but no orange spots. The hindwings are often flushed blue towards the body. The main food plant is Red Clover, with other clovers sometimes used.

FLIGHT PERIOD In northern Europe normally flies in one generation only, early June–early August; further south there are two generations.

HABITAT AND DISTRIBUTION Can occur in a wide variety of flowery grasslands, from damp hay meadows to dry downlands and woodland clearings. Extinct in Britain for more than a century, but still locally common in most of continental Europe except northern Scandinavia.

SIMILAR SPECIES Much larger than the Small Blue (see p. 72), which has similar undersides; much browner below than the Holly Blue (see p. 74). The Green-underside Blue (see p. 77) is usually greyer below, with more blue-green flushing and larger forewing black spots.

Mazarine Blue resting after rain.

Common Blue

Polyommatus icarus

Wingspan 35mm. Easily the most common and widespread of our blue butterflies, this lovely little insect can be found almost everywhere. Males are unmarked bright blue above, with a clear white fringe around the wings; females are rather variable, with a ground colour of brown or bluish, edged with orange dots that fade towards the front of the forewing (they look rather like the male Brown Argus). Below, they have the characteristic undersides of this group, with a brownish-grey base colour dotted with white-ringed black spots, and a continuous wavy orange margin. On the forewings there are two dots close to the body, which distinguish it from the Brown Argus and Silver-studded Blue (see pp. 82 and 80). In northern Scotland and western Ireland, individuals are generally larger and females are more brightly coloured than elsewhere in Britain. The main food plant is Bird's-foot Trefoil, with various other trefoils also used.

FLIGHT PERIOD Most commonly flies in two generations, May–June, then July–September, with an occasional small third generation; in its most northerly sites it has just one brood, July–August.

Female Common Blue covered in dew after overnight roost.

An unusual female Common Blue, typical of populations in Ireland and West Scotland.

Male Common Blue basking.

HABITAT AND DISTRIBUTION Uses a wide range of habitats, wherever its primary food plant occurs, including downland, meadows, pastures, woodland clearings, gardens and wasteground. Occurs throughout Britain except on high mountains and the northernmost islands. Within northern mainland Europe absent only from Iceland.

SIMILAR SPECIES The Brown Argus and Silver-studded Blue are discussed above; male Adonis Blues (see p. 90) are usually brighter turquoise-blue and have distinct chequered wing fringes.

Adonis Blue

Polyommatus bellargus

Wingspan 30–40mm. This is one of the most beautiful of our small butterflies, and the shiny bright turquoise-blue wings of the male stand out clearly from those of other species. Males are all blue above, with a bright white fringe that is chequered with black; females are chocolate-brown dusted with blue scales, and there are a few spots in a patch of orange and blue at the rear of the hindwing. The underwings are very similar to those of the Common Blue (see p. 88), although the distinctive chequered fringes are also visible below. The primary food plant is Horseshoe Vetch, although it will occasionally use Crown Vetch and other legumes in some areas. It is very active in sunshine, but rarely moves far from its clearly defined colonies.

FLIGHT PERIOD Regularly two brooded, emerging first in early May and lasting to mid-June, with a second, often more abundant generation in August and September.

HABITAT AND DISTRIBUTION Within Britain confined to warm, generally south-facing, sloping chalk and limestone downland, usually closely grazed, where the food plant is at its most abundant. A strongly southern species, reaching no further north than the Chilterns and Cotswolds, and occurring in defined colonies. Elsewhere in Europe most common in the south, reaching as far north as northern Germany and the southern Baltic states.

Adonis Blues feeding on Marjoram.

Spring generation female Adonis Blue basking.

Male Adonis Blue, second generation.

SIMILAR SPECIES The Common Blue (see above); female Chalkhill Blues (see p. 92) are almost indistinguishable, differing only in that the spots on the margins of the upperwings are edged with white in Chalkhill, and blue in Adonis, although with practice they can be told apart by colour differences.

Chalkhill Blue

Polyommatus coridon

Wingspan 34–40mm. A strikingly beautiful, rather large blue butterfly. Males are a distinctive pale silvery-blue above, edged with a variable amount of black, especially on the forewings. Females are chocolate-brown, very similar to Adonis Blue females (for differences, see p. 90), although they are slightly larger. Both sexes have chequered black-and-white fringes to the wings. As for the Adonis Blue, the main food plant is Horseshoe Vetch. The Chalkhill Blue is not quite as fussy about the habitat conditions as the Adonis Blue, so it occurs more widely, in more sites. In good years on good sites, it may be very abundant.

FLIGHT PERIOD Flies in a single generation, early July–mid-September.

HABITAT AND DISTRIBUTION Within the UK virtually confined to chalk and limestone downs in England, where the food plant occurs, reaching as far north as Cambridge in the east, and the Cotswolds in the west. Absent from Ireland and Scotland. In northern mainland Europe reaches as far north as northern Germany, although it is absent from many areas.

Chalkhill Blue larva tended by ants.

Chalkhill Blues, female on left, male on right.

Male Chalkhill Blues on Carline Thistle.

SIMILAR SPECIES Males unlikely to be confused with anything else. The chequered fringe separates females from other similar species, except the Adonis Blue (see above).

BRUSH-FOOTED OR FOUR-FOOTED BUTTERFLIES, NYMPHALIDAE

The Nymphalidae is a huge and very varied family that now includes the fritillaries, vanessids and browns. All of these butterflies have vestigial front legs, which are not used for walking.

Fritillaries

The fritillaries are a group of largely orange-brown butterflies that are variously patterned with different shades of black, white and brown.

Dark Green Fritillary

Argynnis aglaja

Dark Green Fritillary feeding on Knapweed.

Wingspan 58mm (smallest males), to 69mm (largest females). The most widespread of our larger fritillaries, with a versatile ability to use different habitats. These are orange-brown butterflies chequered with black. Males are stronger orange-brown than females, with no obvious scent brands and rather rounded wings; females are paler yellowish-brown, although often dark towards the body, with pale spots around the margins of the wings. Underneath, the hindwings are pale straw coloured, heavily flushed with green and decorated with large, oval, pearly-white spots, and narrower white spots around the margin – a lovely combination. This is an active, strong-flying species, although it is not a notable migrant. On warm days the males spend much of their time searching for females and interacting with other males, in between bouts of nectaring on flowers. Females are less conspicuous, seeking out violets, especially Hairy Violet, on which to lay eggs.

Female Dark Green Fritillary on Knapweed.

Male Dark Green Fritillary basking.

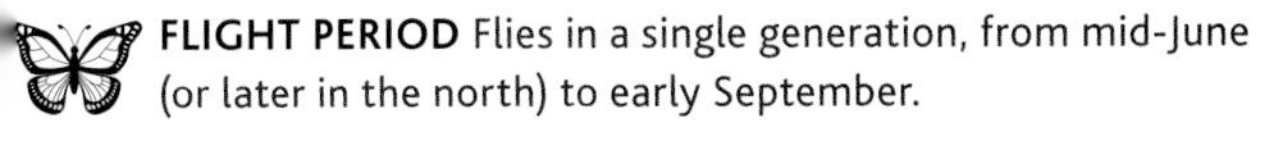

FLIGHT PERIOD Flies in a single generation, from mid-June (or later in the north) to early September.

HABITAT AND DISTRIBUTION Predominantly a species of open, grassy areas, from coastal dunes and moorland edges to warm chalk grassland, throughout the UK as far north as the Orkneys, although generally most common in the west. Locally common in northern mainland Europe except the far north.

SIMILAR SPECIES The High Brown Fritillary and Niobe Fritillary (see pp. 96 and 101) are both similar.

High Brown Fritillary

Argynnis adippe

High Brown Fritillary larva on bracken.

Wingspan 60mm. A beautiful butterfly, very similar in overall appearance to the Dark Green Fritillary (see p. 94), although much rarer and more elusive. From being widespread it has declined so severely in recent decades that it has become one of our rarest species. It is virtually the same size as Dark Green, or perhaps slightly smaller on average. From above, males look almost identical, except that they have thicker sex brands on the second and third veins of the forewing. Female High Browns resemble Dark Greens, except that the pale spots around the margins of both wings are absent in High Browns. Underneath they are similar, but High Browns are generally less greenish, and have a series of small, red-ringed white dots between the outermost two lines of dots on the hindwing, absent in Dark Greens. In general, Dark Greens have more and brighter white spots on the forewing margin, although this varies.

FLIGHT PERIOD Flies in a single generation, mid-June–early September.

HABITAT AND DISTRIBUTION Formerly widespread as a woodland butterfly, but now occurs within the UK in more open, mixed habitats that include grassland, scrub and woodland elements, and frequently Bracken, notably on Dartmoor and the limestones around Morecambe Bay. Confined to these areas and a tiny number of sites in the Midlands and South Wales. Still quite widespread in northern mainland Europe, although absent from many lowland areas and the far north of Scandinavia.

SIMILAR SPECIES Dark Green Fritillary (see above) and Niobe Fritillary (see p. 101).

Male High Brown Fritillary on bramble.

High Brown Fritillary on thistle, showing underwing pattern.

Silver-washed Fritillary

Argynnis paphia

Wingspan 70mm (smallest males), to 80mm (largest females). This strikingly beautiful butterfly is easily our largest fritillary, and one of our largest butterflies. Males are particularly distinctive and conspicuous, with their bright orange-brown ground colour chequered with numerous large black dots and streaks. On the forewings there are four broad black streaks that follow the veins – these are the scent brands, or androconial organs, which open to shower the female with scented scales during courtship. Females, although larger, have a duller, slightly greenish-brown ground colour and are less conspicuous, and they lack the distinctive sex brands. A small proportion of females in some areas have a much duller, pale khaki-green background; they are known as var. *valezina*. The undersides are similar in both sexes, brown washed with green, with four variable nebulous, silvery-white lines running across them – a highly distinctive pattern among northern European butterflies.

These are strong-flying, mobile butterflies; they are not migratory, but are likely to easily find suitable new habitat, so they spread quite quickly under favourable conditions. Their egg-

Male (right) and female Silver-Washed Fritillaries feeding on bramble.

A mating pair of Silver-washed Fritillaries, showing the undersides.

laying behaviour is interesting; the female flutters along low over the ground in search of the food plant, Common Dog-violet or its close relatives. When she finds a suitable violet plant she flies to the base of the nearest appropriate tree and lays the eggs on the

The *valezina* form of Silver-washed Fritillary.

A male Silver-washed Fritillary in flight.

bark, or sometimes on nearby vegetation. The newly emerged tiny caterpillar then hibernates on the tree, and finds its way to the violets when it emerges in spring.

FLIGHT PERIOD Flies in one generation, from about mid-June through to late August or early September.

HABITAT AND DISTRIBUTION Essentially a woodland butterfly, most at home in the rides and glades of larger deciduous forests, although in western Britain it is more adaptable, often using hedgerows and other less wooded habitats. A western species, found in central and south-west England (with a few northern outliers), Wales and Ireland. Widespread in the rest of northern Europe, although absent from Denmark and northern Scandinavia.

SIMILAR SPECIES Although other large fritillaries are quite similar, the undersides and the bold scent brands of the male Silver-washed Fritillary are quite distinctive.

Niobe Fritillary

Argynnis niobe

Wingspan 50–60mm. Although this species has been recorded in the UK over the years, it is believed that the records were of introduced specimens, or mistakes, and it is not currently considered to be a UK species. It is remarkably similar to the High Brown Fritillary (see p. 96), and needs close examination, preferably of the underside, to separate it. Males from above are slightly smaller, have a slightly convex forewing margin and do not have the thickened sex brands of High Browns. Females' uppersides are rather duller and greyer than those of the High Brown. Underneath, the red-ringed spots are smaller and more regular in this species, and there is an additional small, yellowish, black-centred spot near the base of the hindwing.

FLIGHT PERIOD Flies in one generation, from about early June through to August.

HABITAT AND DISTRIBUTION Flowery grasslands and clearings where there are violets, especially in uplands, in most of northern Europe except the UK and northern Scandinavia.

SIMILAR SPECIES High Brown Fritillary (see above).

A newly-emerged Niobe Fritillary on thrift.

Queen of Spain Fritillary

Issoria lathonia

Wingspan 38mm (smallest males), to 55mm (largest females). An attractive medium-sized fritillary. Males are tawny-orange above with the typical fritillary dark markings, mainly round spots although with a few lines. Females are larger and slightly duller, with a larger area of greenish-grey on the hindwing bases. With practice, it is possible to identify this species from above both by its size and by the slightly concave margin to the forewing, which can be surprisingly noticeable. Underneath, this is a striking butterfly, with both sexes being similar; on the hindwing there are six or so large, silver-white patches on the half closest to the body, then a red line with white-eyed dark spots in it, and finally a curve of large white spots on the margin. This is a highly mobile migratory butterfly, constantly moving outwards from its southern base, with evidence of some reverse migration as well. Its larvae feed on Field and Wild Pansies.

Queen of Spain Fritillary on thistles.

FLIGHT PERIOD In southern Europe almost continuously brooded March–October, and numbers tend to build up by late summer. In good years there is outwards migration northwards, with most records in the UK being in late summer.

HABITAT AND DISTRIBUTION Because it is so mobile, it can turn up in almost any flowery habitat, although pansies are required for breeding colonies to establish. Common over much of central and southern Europe, but an increasingly rare migrant northwards to central England and southern Scandinavia. Occasional recent records of breeding in the UK.

SIMILAR SPECIES The concave forewings and large white patches should separate it from other fritillaries. Can slightly resemble a Comma (see p. 130), but the deeply incised wing margins of that species are distinctive.

Queen of Spain on scabios, showing the distinctive underside.

Small Pearl-bordered Fritillary

Boloria selene

Wingspan 37mm (males), to 45mm (females). A most attractive little butterfly, once common throughout much of Britain but now, like many other fritillaries, a declining and uncommon species. Both sexes have the typical tawny-orange upperwings, chequered with black dots and stripes. Apart from in size, the sexes are rather similar, although females may be slightly duller than males, and often (although not always) have pale spots around the margins of all four wings. The underside hindwings are more distinctive; they are similar in both sexes, with seven silvery spots around the margin ('pearl border'), and 6–7 other large white patches within a matrix of cream and orange-yellow patches. The butterflies are active in warm weather, although they rarely fly far and have poor powers of colonisation unless population numbers are high. The main food plants are Common Dog-violet and Marsh Violet.

FLIGHT PERIOD Normally single brooded, on the wing mid-May–early July, although may occasionally have a small second brood in August (and regularly does so further south in mainland Europe).

Male Small Pearl-bordered Fritillary on thistles.

Female Small Pearl-bordered Fritillary basking.

HABITAT AND DISTRIBUTION In the UK occurs in a variety of habitats, especially in the west, where it is most common – cliffs, coastal grassland, moorland edges, woodland clearings, all generally rather damp; much rarer further east and confined more to woodland. Widespread in mainland northern Europe, extending well beyond the Arctic Circle.

SIMILAR SPECIES The Pearl-bordered Fritillary is very similar (for differences, see p. 106).

Male Small Pearl-bordered Fritillary showing the distinctive underwings.

Pearl-bordered Fritillary

Boloria euphrosyne

Pearl-bordered Fritillary on Violet leaf.

Wingspan 38–47mm. As the name suggests, this species is slightly larger than the Small Pearl-bordered Fritillary (see p. 104), with a larger wingspan, although there is not enough of a size difference between the species to identify it by. From above the two species are very similar, but there are a couple of minor consistent differences. In this species the marginal dark chevrons are not attached to the outer wing-margin dark colour – they are floating in orange – and the next line of spots in is roughly midway between the chevrons and the further line of dark markings. On the underwings the differences are more clear cut, as this species just has two large white patches (not 6–7) in addition to the marginal pearls. The food

Male Pearl-bordered Fritillary on hawkbit flower.

Pearl-bordered Fritillaries feeding.

plants and behaviour are similar in both species, although the earlier emergence of this species can be a useful clue to identification.

FLIGHT PERIOD Normally single brooded, on the wing from late April (or even earlier nowadays). It may occasionally have a small second brood in August, and regularly does so further south in mainland Europe.

HABITAT AND DISTRIBUTION Occurs in slightly drier sites than Small Pearl-bordered, in woodland rides and clearings, moorland edges, grassland with bracken and scrubby coastal valleys. It has declined in the UK even more than Small Pearl-bordered, and is confined to relatively few areas in western and northern Britain, with some sizeable populations in the Cairngorms area. Still quite common in much of mainland northern Europe, as far north as northern Scandinavia.

SIMILAR SPECIES Small Pearl-bordered Fritillary (see above).

Marsh Fritillary

Euphydryas aurinia

Wingspan 40–50mm. A highly variable but distinctive little variegated butterfly, with males generally smaller than females. The sexes are otherwise broadly similar, and both have an interesting and attractive multicoloured pattern on the upper surfaces, with a mixture of orange, khaki-yellow and black markings in a longitudinally stripey pattern. They also have distinctive black dots in the orange band towards the rear of the hindwing, and no dots on the forewings. Underneath, the hindwings are rather similar, although lacking the black markings; unlike many other small fritillaries, they have no white markings below. They are not particularly conspicuous butterflies. Males patrol low over vegetation looking for females, quickly settling if the temperature cools, while females fly around quietly looking for egg-laying sites. The predominant food plant is Devil's-bit Scabious in damp or dry localities.

FLIGHT PERIOD Invariably flies in a single generation, emerging in mid-May and eventually disappearing by mid-July.

Female Marsh Fritillary egg-laying on Devil's-bit.

Male and female Marsh Fritillaries feeding on Bugle.

HABITAT AND DISTRIBUTION Occurs in two quite different habitat types, linked by the presence of the food plant. The most frequent habitat is wet, sedge-rich meadows in western Britain and Ireland, where the scabious grows large and abundantly. Relatively recently, probably because of a decline in sheep

Female Marsh Fritillary feeding on Kingcup.

Knapweed Fritillary basking.

grazing, the butterfly has colonised dry chalk and limestone downlands in western England, from Dorset to Gloucestershire, and is doing well there, in contrast to its steady decline in wet meadows as more and more of them have been drained. In northern mainland Europe it has a very patchy distribution; it is common in places such as north-west France, yet absent from many lowland areas and most of Scandinavia. Populations vary cyclically within these wider ranges, due to the rise and fall of populations of a parasitic wasp.

SIMILAR SPECIES The Knapweed Fritillary *Melitaea phoebe* can look very similar from above because of its rather similar variegated pattern. However, it lacks the black dots in the orange cells around the hindwings which the Marsh Fritillary has; underneath, it is much paler, with bands of near white, and it lacks the same submarginal black dots on the undersurface. It is a common species in southern Europe, occurring in flowery meadows wherever knapweeds grow, and extends northwards as far as northern France, central Germany and the Baltic states. It is absent from the UK and Scandinavia.

Meadow Fritillary

Melitaea parthenoides

Wingspan 32–36mm. A tiny fritillary – one of the smallest in the area. The upperwings are the typical fritillary orange-brown colour, but marked more with lines and squares, and with virtually no spots. Underneath, there are marginal and central white bands on the hindwing, which also lacks any spots. The food plants are Ribwort and Hoary Plantain.

FLIGHT PERIOD Commonly two broods, May–June, then August–September, although often single brooded in midsummer in mountain areas up to 2,000m.

HABITAT AND DISTRIBUTION Mainly meadows and other flowery grasslands, often damp, including mountain areas. A south-west European endemic from northern France southwards.

SIMILAR SPECIES The Glanville Fritillary (see p. 112) is similar but larger, with black spots in the white areas on the under hindwings. The Heath Fritillary (see p. 114) can look very similar, but is larger, and usually with more black above and whiter below; there is also a broad line at the mid-point of the rear of the upperside forewing that is oblique in Meadow but vertical in Heath.

Meadow Fritillary basking.

Meadow Fritillary roosting.

Glanville Fritillary

Melitaea cinxia

Wingspan 40–50mm. An extremely attractive little fritillary, generally similar in colour and pattern to other small fritillaries, and with females being larger on average than males. Apart from this the sexes are very similar, and their best distinguishing feature from above is the row of black dots in submarginal orange cells around the rear of the hindwings, but *not* on the forewings, a characteristic shared only by the Marsh Fritillary (see p. 108) in the UK. The underside is quite distinctive, with two bands of orange interspersed with three bands of white or cream, of which the two outermost have black dots in every cell. In good weather the butterflies are active and strong fliers, with males moving constantly in search of females or nectar. The primary food plant is Ribwort Plantain, or Buckshorn Plantain in extremis.

FLIGHT PERIOD Single brooded throughout its northern European range, flying May–June, although further south and in exceptional years it has two broods.

Glanville Fritillary on Lady Orchid showing undersides.

Female Glanville Fritillary on Ribwort Plantain.

Male Glanville Fritillary on Bugle flowers.

HABITAT AND DISTRIBUTION In Britain this is a rare butterfly restricted to unstable coastal areas on the Channel Islands, Isle of Wight and a few adjacent coasts where the food plant is abundant. Elsewhere in northern Europe it is reasonably common in grassland, clearings and scrub, as far north as southern Scandinavia and Estonia.

SIMILAR SPECIES The Marsh Fritillary is much more variegated in colour above, and on the underside there is just one line of black dots in the orange submarginal cells.

Heath Fritillary

Melitaea athalia

Male Heath Fritillary on Bugle.

Wingspan 40–46mm. A typical small fritillary with the usual upperside mixture of orange background with black markings. The sexes are similar, except that males are generally smaller and slightly darker than females above, and overall this species tends to look darker, with more black, than other similar species. There are no black spots at all in any orange cells, which distinguishes it from all the other UK small fritillaries (but see Meadow Fritillary, p. 111). Underneath, it looks similar to other species, but again there are no black spots in either the orange or cream cells. Its main food plant in the UK is the annual Cow-wheat, although it does use speedwells, Foxglove and plantains, and elsewhere in its range it readily uses any of these food plants.

FLIGHT PERIOD Normally single brooded, about mid-May–early August, although in good years and warmer areas there is a small second brood, late August–September.

Female Heath Fritillary roosting.

HABITAT AND DISTRIBUTION In the UK, a very restricted distribution in southern England, only in a few woods and moorland valleys where the food plant is abundant. Widespread virtually throughout northern mainland Europe, occurring in a much wider range of grassy, flowery habitats.

SIMILAR SPECIES The False Heath Fritillary *M. diamina* is very similar, but generally has darker hindwing uppersides, and below there are usually some black dots and the hindwing margin is yellow. Occurs in scattered, grassy localities, often montane, in central to northern Scandinavia. See also Meadow Fritillary (above).

Aristocrats

Within the Nymphalidae there is a distinctive group of butterflies known as the aristocrats. They are generally large and colourful species, often with indented wing margins, and have well-protected, spiny caterpillars.

Red Admiral

Vanessa atalanta

Wingspan 65–75mm. One of the most common and distinctive of our butterflies, and arguably one of the most beautiful. The sexes are very similar, although males are slightly smaller than females. The uppersides are a beautiful velvety dark brown, with scarlet edges to the hindwings, a bold scarlet band across the forewings and large white patches towards the forewing tips. Underneath, the forewings are as boldly marked as the uppersides, but the hindwing is subtly marbled with black, brown and blue, and only this is visible when the insect is fully closed, resulting in it becoming remarkably well camouflaged.

These are habitual migrant butterflies whose base is in the Mediterranean area, from where they migrate outwards each year, often reaching well into Arctic regions. A lot of work has

Red Admiral larva on nettle.

Red Admiral feeding on thistle, showing both underwings.

Red Admiral basking on dead leaves in autumn.

been carried out on their migration and survival patterns, which demonstrate an extraordinary story. Many individuals begin migrating northwards in spring, and numbers gradually build up throughout the summer. Populations become established in northern areas as the arrivals breed, but as temperatures fall they begin to migrate southwards in vast numbers; many have been shown to travel virtually non-stop from northern Europe back to the Mediterranean over a period of 2–3 weeks. Individuals do not normally survive northern winters – although they may do so occasionally – and it seems that genuine hibernation is rare even in southerly areas.

FLIGHT PERIOD Seen in any month of the year in Britain, although numbers in winter are very small and may consist of either individuals surviving through a warm winter, or erratic migrants. At the time of writing (January 2014) individuals are being frequently seen in southern England, in an exceptionally mild winter. Peak numbers are usually seen July–September.

Red Admiral feeding in garden on Sedum flowers.

HABITAT AND DISTRIBUTION Red Admirals are so mobile that they have no specific habitat as such. Their food plant is Stinging Nettle, which is common almost everywhere, and they have a continuing need for nectar or, in autumn, the juice of ripe fruits, so they are common in gardens, parks, woodland clearings and other warm, flower-rich sites. They arrive in variable numbers throughout the UK and as far north as northern Scandinavia, and even Iceland.

SIMILAR SPECIES No common species can be confused with it. The Indian Red Admiral *V. indica* is similar, but has a larger area of scarlet on the forewing with patches of black within it. It turns up occasionally, probably as an escapee rather than a migrant.

Painted Lady

Vanessa cardui

Painted lady larva on thistle leaves.

Wingspan 62–72mm. A most beautiful butterfly, as befits its name, and also a highly successful one, occurring on every continent except Antarctica and South America. The sexes are similar, although males are smaller on average than females. The upperwings have a base colour of orange to pale pink (which fades with age), patterned with black, and with extensive black tips marked with white. Underneath, the forewing is boldly patterned, but the hindwing is an intricate pattern of colours, lines and spots. In northern Europe this species occurs as an annual migrant only, although the butterflies breed on arrival, using a variety of thistle species, but do not survive the winters in any form. Recent research has found that huge numbers migrate southwards each autumn, flying too high to be readily noticed, at heights of up to 1,000m.

FLIGHT PERIOD Breeds in late winter in North Africa, and in wet winters is produced in large numbers. Many of these move northwards, arriving in Britain and northern mainland Europe from April onwards.

HABITAT AND DISTRIBUTION Neither as consistent nor as abundant as Red Admirals (see p. 116), although in some years, such as 2009, it arrives in huge numbers – in the UK 28 million butterflies were estimated to have arrived in one weekend in late May 2009. In good years it spreads throughout Britain and the whole of northern mainland Europe, and may turn up anywhere flowery.

SIMILAR SPECIES The American Painted Lady *V. virginiensis* differs in having large blue eye-spots on the upper hindwings, and two large eye-spots below. It is a very rare natural migrant to the UK.

Painted Lady resting whilst migrating over a high pass.

Painted Lady on Sea-holly.

Peacock

Inachis io

Wingspan 64mm (smallest males), to 74mm (largest females). A spectacularly beautiful and distinctive butterfly which, fortunately, is common and widespread throughout the area. The sexes are virtually identical in markings, although females are larger on average than males. The ground colour is a rich reddish-brown, and each wing has a large and conspicuous 'eye'. The eyes on the hindwings are round, blue and edged with cream, while those on the forewings are more oval and blurred, with mixed blue, yellow, brown and black. All wings have wavy-edged, dark greyish margins. Underneath, the hindwings – the area usually seen when the wings are fully closed – are dark charcoal-grey to black, marked with a few lines and darker or lighter patches, so that a closed individual looks remarkably like a leaf.

Peacock feeding on bramble, showing the dark underwings.

Above: Peacock feeding on Hemp Agrimony flowers.

Left: Red admiral larvae on nettle leaves.

Although this is a strong flier and quite mobile in search of new habitats, it is not a migrant like the Painted Lady or Red Admiral (see p. 120 and 116). Instead, the adults habitually hibernate, and they live for almost a year. In late autumn, as the weather cools, they seek out potential hibernation sites – hollow trees, sheds, caves, crevices and occasionally houses – where they pass the winter, although they do emerge on warm days in any month. This is a highly vulnerable period, and relatively few survive the winters. Apart from the camouflage, they have defensive mechanisms such as flicking and hissing, making them seem more like predators than prey. Their food plant is almost invariably Stinging Nettle, although occasionally Hop or Wall Pellitory is used.

Peacock in hibernation in a shed.

Close view of the antennae of a Peacock.

FLIGHT PERIOD Newly emerged adults first appear in July, and these then survive through to the following June. They may appear in any month, but are most likely to be seen April–June, and July–October.

HABITAT AND DISTRIBUTION Does not occur in sedentary colonies, but flies widely in search of mates, food plants and food, so can turn up anywhere, but is most common in and around deciduous woodland. One of the most common garden visitors. Widespread in Britain except the Highlands, and northwards to central Scandinavia. Since the 1970s it has extended its range considerably northwards into Scotland and Scandinavia, with a consequent ability to hibernate through the longer, colder winters.

SIMILAR SPECIES None.

Small Tortoiseshell

Aglais urticae

Wingspan 46–58mm. One of our most familiar butterflies, this is perhaps *the* most common garden visitor of all the butterflies. Males and females are identically marked, and males are slightly smaller than females. The uppersides are a lovely pattern of colours, essentially tawny-orange, with a chequered front margin comprising alternating black and yellow with a bit of white. Around the wavy margins of both wings there is a continuous line of pearly blue, crescent-shaped spots. Underneath, the forewing is straw coloured with darker markings, while the hindwing is dull brownish-black, with a paler area towards the hind margin and an edge of inconspicuous blue spots. When the wings are fully closed, the butterflies look like dead leaves.

Although this is quite a mobile species, and adept at colonising new territory, it is not a habitual migrant like the Red Admiral (see p. 116). The late-summer generation hibernates, seeking out suitable places, and the species is much more attracted to houses, barns and churches – and other man-made sites – than is the

Small Tortoiseshell larva on nettle leaf.

Small Tortoiseshell feeding on thistle, showing underwings.

Small Tortoiseshell feeding on Buddleia in a garden.

Peacock (see p. 122). On emergence in spring the males become quite territorial, setting themselves up in suitable nettle patches to seek out or attract passing virgin females. They have quite a distinct diurnal rhythm, passing the morning in active feeding or basking, depending on the weather, and the afternoon in territorial displays and courtship. In recent years the species' numbers have declined sharply – a 74 per cent decline was estimated for the decades since 1976 in the UK, and the shocking summer of 2012 saw them at their lowest ebb. Fortunately, numbers seem to have recovered considerably after the warm late summer of 2013. It is not entirely certain why they declined so much, but it is probably because of a combination of a series of poor summers, as well as the advent of a new parasite, the fly *Sturmia bella*, which has invaded from continental Europe in recent years. Populations have always fluctuated in response to weather and other parasites, so there is hope that numbers may recover again. The food plant is Stinging Nettle, or less commonly Annual Nettle.

FLIGHT PERIOD Adults emerge from spring, in around April, depending on the weather, and they soon mate and die off. The next generation emerges in June through July, and a second generation normally emerges in August. This generation then hibernates. Adults may be seen in any month if the weather is warm.

HABITAT AND DISTRIBUTION Flies widely in search of mates, food plants and food, so can turn up anywhere, but is most common where there are lush nettle patches in sunshine, and in areas where there are abundant nectar sources. Gardens and parks are especially favoured, and the butterflies have a special liking for Buddleias and the stonecrop *Sedum spectabile*.

SIMILAR SPECIES The Large Tortoiseshell (see p. 132) has similar markings, but is much larger, rather duller brown in colour and has no white on the forewings.

Small Tortoiseshells hibernating in a garden shed.

Masses of Small Tortoiseshells feeding on Sedum spectabile in a garden.

Comma

Polygonia c-album

Comma larva on nettle leaves.

Wingspan 52–62mm. This highly distinctive and attractive butterfly is now once more a familiar sight after a long period of decline. The sexes are broadly similar, both tawny-orange with scattered dark markings and a few lighter ones, and markedly jagged edges that are edged with brown. Males are slightly smaller than females. They have more jagged edges than females, and are a deeper orange colour. Underneath, the butterflies are marbled pale brown and dark brown, with the jagged wing edges still very noticeable, and here is the feature that gives the species its name – in the centre of the hindwing there is a distinct white comma shape. The earliest adults to emerge are usually much more golden and bright in colour than later ones; they are known as form *hutchinsonii*, and they quickly mate, producing a late-summer generation.

From being a common species in the 19th century, the Comma declined to very low levels in the early part of the 20th century, but has since recovered strongly both in population numbers and range. The decline was probably due to the loss of hop fields (Hops were its main food plant) and poor weather; the expansion is likely due to warming weather and its transfer to nettles as the primary food plant. Hops are still used occasionally, as are elms.

FLIGHT PERIOD The first generation of the year appears in early June, and is followed by a second generation in late August. The autumn generation hibernates, usually in woodland, and reappears in early spring, although like many hibernators the butterfly may be seen in any month.

HABITAT AND DISTRIBUTION Although Commas are not migrants, they are quite mobile and can turn up almost anywhere, but are most common in and around the sunnier parts of woods.

Comma, male *hutchinsonii* form basking.

Newly-emerged Comma female in midsummer.

They are widespread and common in England and Wales, rare in Scotland and virtually absent from Ireland. They have spread in northern mainland Europe, and now reach into northern Scandinavia.

SIMILAR SPECIES Although Commas are superficially similar to fritillaries in colour (especially in form *hutchinsonii*), their jagged edges and undersides are quite different.

Large Tortoiseshell

Nymphalis polychloros

Wingspan 65–74mm. A large and beautiful butterfly. Males and females are virtually identical, except that females are slightly larger on average. The species is quite similar to the Small Tortoiseshell (see p. 126), but is substantially larger and a duller brownish-orange, and lacks any white on the forewings. In addition, there is much less black on the upperside hindwings, and the pretty blue marginal forewing spots of Small are almost invariably absent. Underneath, the butterflies are dark brown marbled with lighter brown, looking – like many of this group – very like dead leaves when they are closed.

Once quite a common resident, by the 1950s this species had declined in the UK to the point of extinction for a complex of reasons. Since then there have been regular sightings, but most are assumed to have been of accidental or deliberate introductions. However, there has been a steady trickle of recent records from the Isle of Wight, including of overwintered individuals, raising the possibility that it may come back as a resident. The food plants are elms (its main food plant in the UK), Wild Cherry, Aspen and willows.

FLIGHT PERIOD Adults hibernate, and appear in spring, in late March or early April, then die soon after mating. The new generation appears in late July or early August, and goes into hibernation quite soon afterwards. Large Tortoiseshells are thus not on the wing for very long.

HABITAT AND DISTRIBUTION Essentially an insect of sheltered deciduous woodland, where the food plants occur, although it is quite mobile. Records in the UK are normally from England. In northern mainland Europe it is widespread and locally common as far north as southern Scandinavia and Finland.

SIMILAR SPECIES Small Tortoiseshell (see above).

Male Large Tortoiseshell basking.

Female Large Tortoiseshell.

Camberwell Beauty

Nymphalis antiopa

Wingspan to 88mm. This large and spectacular butterfly is made even more special by virtue of its extreme rarity, in the UK at least. The sexes are similar (although females are slightly larger than males), and quite unmistakable. The ground colour is a rich reddish-brown when fresh, ageing to dark chocolate-brown, edged with a conspicuous broad white, creamy or yellow border. Between the border and the rest of the wing there is a line of blue spots on each wing. The underwings are very dark, except for the pale border. This is a very strong-flying, mobile butterfly, and it is believed that all UK sightings are of immigrants (or released specimens), and not of individuals that are resident. When it does migrate, it may do so in groups, and quite large numbers can appear in some years. It is a hibernator, passing many months in hibernation. The food plants include willows, poplars and birches.

FLIGHT PERIOD New-generation individuals appear in July or even August, fly for a short while, then move into hibernation; they reappear in spring in April, when they mate and continue flying until June.

Newly-emerged Camberwell Beauty next to empty pupal case.

HABITAT AND DISTRIBUTION A very mobile butterfly that can turn up almost anywhere, but has a particular liking for large deciduous woods and river valleys. In Britain the species is an occasional migrant, mainly to eastern England; elsewhere in northern Europe it is widespread and quite common as far north as northern Scandinavia.

SIMILAR SPECIES None.

Newly-emerged Camberwell Beauty basking.

An overwintered Camberwell Beauty showing typical worn wings.

Map Butterfly
Araschnia levana

Wingspan 30–48mm. This is an extraordinary and beautiful little butterfly, with one rather unusual feature – its two generations are totally different in appearance, so much so that they were originally thought to be different species, and still have separate form names now. Males and females are similar, except that females are generally larger.

The spring form, known as *levana*, is rather like a small fritillary. The upperside ground colour is orange well marked with black spots, and with black margins and wing-tips, as well as a line of thin blue spots along the hindwing margins. There are scattered white spots, especially towards the tips of the forewings. The undersides are strikingly beautiful, a mosaic of brown, purple, blue, white and cream, with white lines, forming the 'map' of the common name. The summer form, *prorsa*, is quite different, most resembling a small White Admiral (see p. 140). The upperside ground colour above is dark bluish-black,

Female spring form, *levana* of Map Butterfly.

with a thin, white interrupted margin, a bold arc of virtually contiguous white spots across the middle of both forewings and hindwings, and an interrupted thin streak of red just inside the margin of the hindwing. Underneath, it is rather similar to the spring generation, although with a bolder white streak across, and less blue. Recent research has shown that day length and development time control the appearance of the different forms, although as yet it is not understood what advantages it confers on the species, as no other butterfly in Europe shows this kind of dimorphism.

The butterfly's main food plant is Stinging Nettle (with other nettles used less often), on which the female lays very distinctive columns of ten or so eggs, in little clusters of columns, on the undersides of the leaves.

FLIGHT PERIOD *Levana* flies late April–June; summer generation *prorsa* flies July–August. The second brood is usually much more numerous than the first. Occasionally there may be a small third brood.

Underside of spring form of Map Butterfly, showing the 'map'.

Dark Spring form of Map Butterfly.

HABITAT AND DISTRIBUTION Occurs in quite a wide range of warm, sheltered habitats, especially woodland clearings and margins, along hedges, in scrubby grassland and even in gardens. Not particularly mobile, and is found mainly in areas where there are nettles.

Widespread and locally common in central and eastern Europe, absent from the Mediterranean area, and found no further north than northern Scandinavia. It has been steadily spreading northwards in recent decades and has now reached central Finland. In Britain there are occasional records, which are probably mainly releases, but colonisation is not unlikely in the near future.

SIMILAR SPECIES The spring generation looks superficially like a little fritillary, but the detailed pattern, especially on the underside, is quite different. The summer generation has the pattern of a White Admiral (see above), but is distinctly smaller, and has a much more marked red line on the hindwings and quite different undersides.

Summer generation, form *prorsa* of Map Butterfly, female, on marjoram.

Underside of summer generation Map Butterfly.

White Admiral

Limenitis camilla

Wingspan 56–65mm. A beautiful and elegant butterfly redolent of high summer and flowery woodland glades. Males and females are very similar in their colour and markings, although females are larger. The uppersides are dark blackish or brownish-black, with a broad and discontinuous, conspicuous white band across both wings. There may be small patches of orange-red towards the rear of the hindwings, but these are not always visible. Underneath, the wings are strikingly beautiful, a complex pattern of bright orange-brown, the same white band, and a mixture of black spots and grey areas.

The primary food plant is Common Honeysuckle, although other species of honeysuckle such as Fly Honeysuckle and the related Snowberry may be used. Females do not select robust, open-grown plants, but prefer the more attenuated, semi-shaded individuals; thus, unlike most other butterflies, the White Admiral is more associated with the shadier parts of woods. White Admirals (especially males) are strong fliers, with a distinctive rapid flight that incorporates episodes of gliding; several close

The distinctive and attractive larva of White Admiral.

White Admiral settled, showing the beautiful underside.

Male White Admiral feeding on bramble flowers.

The striking pupa of White Admiral.

relatives in mainland Europe are known as 'gliders'. They can be difficult to get close to, as they spend much of their time flying, feeding and basking up in the tree canopy.

FLIGHT PERIOD Normally flies in one generation, mid-June–mid-August, although there may be a small second brood in September in warmer years.

HABITAT AND DISTRIBUTION Essentially a species of large areas of woodland, especially deciduous woodland, although coniferous woods with patches of broadleaved trees will do. In the UK virtually confined to southern and eastern England, although the last few decades have seen a gradual increase in range. Common and widespread in central and western mainland Europe, but absent from most of the Mediterranean area and much of Scandinavia.

SIMILAR SPECIES The Southern White Admiral is very similar (for differences, see p. 143), as is the summer form of the Map Butterfly (see p. 136).

Southern White Admiral

Limenitis reducta

Southern White Admiral basking.

Wingspan 50–60mm. A very similar species to the White Admiral (see p. 140), and difficult to tell apart from it by superficial appearance or behaviour. With a good view, the differences become apparent: on the upperside the wings usually have a bluish tint, and there is a clear white spot midway along the front-wing front edge (absent or much reduced in the White Admiral). Underneath, the pattern is broadly similar, except that there is just one row of submarginal black spots in this species, but two rows in the White Admiral, and the ground colour of Southern is more purplish.

FLIGHT PERIOD One or two generations, mid-May–late June, and July–August; at its mainland northern European sites it is generally single brooded, June–July.

HABITAT AND DISTRIBUTION Similar habitats and food plants to those of the White Admiral, although often found in drier sites. A southern European species, reaching northwards to northern France and southern Germany.

SIMILAR SPECIES White Admiral.

Purple Emperor

Apatura iris

A week-old larva of Purple Emperor.

Wingspan 70–85mm. Without doubt, this is one of the most attractive and charismatic of British butterflies, eagerly sought after by naturalists and photographers. It is the second largest British butterfly, with males being smaller than females. Although the sexes are generally similar in pattern, there are some differences between them. Males have a lovely purple sheen, which only becomes visible when the light falls at an angle, so it is rarely all visible at once. As the insect moves, the sheen constantly flickers and flashes in a quite spectacular way. There is a broad white stripe across all wings, uninterrupted on the hindwing but more broken and patchy on the forewing. There is a dark, orange-ringed eye-spot towards the rear of each hindwing. Females are larger

Male Purple Emperor feeding on sap.

Purple Emperor egg on willow leaf.

than males, lack the purple sheen, and have a broader white stripe and slightly bigger eye-spots. The undersides are similar in both sexes – they are beautifully marbled and striped with orange, chestnut, white, dove-grey and black.

Due partly to its beauty and rarity, this butterfly has been well studied, and its life cycle and behaviour are reasonably well known. It occurs at very low densities even on good sites, so a

Female Purple Emperor basking on oak foliage.

Mature final instar Purple Emperor larva.

mechanism has evolved to allow the sexes to meet. From mid-morning onwards males begin to fly extensively around their woodland, gradually gaining height and eventually gravitating towards the trees in the neighbourhood that are on the highest point. Here they set up territory in prominent positions, ready to attack rivals or intruders, or seek out females. The key trees are known as 'master trees', or more accurately as 'master groves', as several trees are involved, and they are likely to be used year after year by successive generations. They may be any species of tree, and are not necessarily particularly old or tall. Although males spend much time in the canopy, they do come down to feed at mineral-rich damp areas, sap runs and droppings, although they never visit

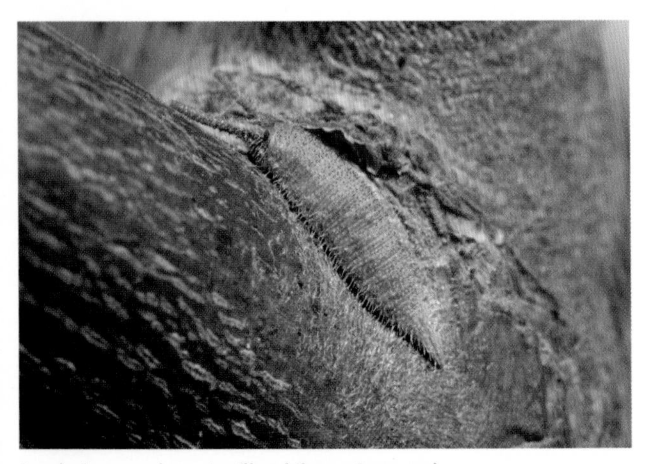
Purple Emperor larva in silken hibernation pouch.

Purple Emperor feeding in damp mud.

flowers; the females assiduously seek out food plants, mainly Goat Willow or other sallows, on which they lay their eggs. It is always worth carrying binoculars when searching for Purple Emperors.

FLIGHT PERIOD Single brooded, on the wing roughly mid-June–mid-August, depending on the locality.

HABITAT AND DISTRIBUTION Primary habitat lies in and around large deciduous or mixed woodland, especially where there are abundant connecting hedges and copses, on heavy soils where Goat Willow is most common. In Britain the stronghold is in central southern England, from Sussex to Wiltshire, with outliers from Devon to the borders of East Anglia, and especially Northants. Widespread and locally common in central mainland Europe, from France to southern Finland, reaching just into southern Sweden.

SIMILAR SPECIES Much larger and bluer than the White Admiral (see p. 140). The Poplar Admiral and Lesser Purple Emperor (see pp. 148 and 149) are similar.

Poplar Admiral

Limenitis populi

Poplar Admiral showing upper wing surface.

Wingspan 72–86mm. Although it could be confused with the purple emperors, this huge butterfly is more closely related to the white admirals. From above it is a striking dark velvety bluish-black, with a bold white stripe across the forewings and an interrupted one across the hindwings. On the rear margins of the hindwings there is an arc of orange-and-brown semi-circles, with a fainter echo on the forewings. Underneath, it closely resembles a large Southern White Admiral (see p. 143), although it is brighter orange. The food plant is Aspen.

FLIGHT PERIOD Flies in a single generation, late May–mid-July.

Poplar admiral underside.

HABITAT AND DISTRIBUTION Mainly in deciduous woods or wooded countryside, where Aspen grows in quantity. Absent from the UK. In continental Europe, widespread but local from France northwards to southern Scandinavia and Finland, and gradually expanding its range.

SIMILAR SPECIES The Purple Emperor and Lesser Purple Emperor (see pp. 144 and 149) are most similar.

Lesser Purple Emperor

Apatura ilia

Male Lesser Purple Emperor on oak leaves.

Wingspan 60–65mm. A smaller version of the Purple Emperor (see p. 144), this is still a large and striking butterfly. The general pattern of the upperwings is similar to the Purple Emperor's; males are beautifully purple sheened, but differ in having a weaker white hindwing stripe, and a more clearly defined, black-ringed-with-orange eye-spot in the margin of the forewing. Underneath, the white hindwing stripe is much less conspicuous than that of the Purple Emperor. Aspens and willows are the main food plants.

FLIGHT PERIOD In northern Europe, mainly single brooded, June–July; two generations elsewhere.

HABITAT AND DISTRIBUTION Woodland, wooded countryside and river valleys, where the key food plants occur.

SIMILAR SPECIES Most similar to the Purple Emperor (see opposite).

Underside of Lesser Purple Emperor.

Browns

The browns, formerly in their own family, the Satyridae, but now a subgroup – Satyrinae – of the larger Nymphalidae, are a group of largely brown butterflies (with occasional exceptions like the marbled whites), frequently with eye-spots on the forewings.

Speckled Wood

Pararge aegeria

Wingspan 46mm (smallest males), to 55mm (largest females). A common and widespread species, although not particularly conspicuous. Males and females are broadly similar in pattern, but there are subtle variations between the sexes and the different generations. The butterflies have a ground colour of deep chocolate-brown; the upperside forewings have six or more large, pale yellow patches, and a single dark, white-centred eye-spot in a yellow patch towards the tip of the wing. The hindwing upper surfaces are darker, with fewer pale patches, and are edged with three eye-spots around the hind margin of the wing. Overall, the yellow patches are smaller on males, and tend to be paler in both

Speckled Wood, ssp. *insula*, Isles of Scilly form.

Female Speckled Wood basking.

sexes of the second generation. In general, this is quite a variable species, with a number of geographically distinct regional variations, such as the larger and paler northern Scotland form, and the more orange Scilly and Channel Islands form. The edges of the wings have a distinctly scalloped appearance.

The life cycle of the Speckled Wood is a fascinating one that has been studied in some detail. Throughout the summer there is a complicated mixture of overlapping generations, with individuals developing at different speeds, and the butterfly has the unique ability (within UK species) to overwinter as a pupa or caterpillar. There is considerable variation in the markings and particularly the darkness or paleness of the ground colour of individuals, especially males, and it seems that this gives different individuals a better chance of succeeding under different conditions. The spring generation is palest, counterintuitively, so that the butterflies warm up relatively slowly – it seems that this is because the weather is much more suited at this time of year to basking and short exploratory flights, whereas in the warm days

Speckled Wood underside.

of summer males fly more widely and patrol their areas. Patrolling is a cooling activity because of the air flow involved, so summer individuals tend to be darker. This variation also exists within generations, giving the population a great ability to adapt to changing weather and local conditions. The species' food plants include Cocksfoot and other lush grasses growing in semi-shaded conditions.

FLIGHT PERIOD In warm years may be seen almost continuously late March–October; in cooler years this time span separates into two main generations, and a limited third one.

HABITAT AND DISTRIBUTION A more shade-tolerant species than most butterflies, so found in woodland, around the edges of glades, along hedges, and in parks and gardens, but rarely far from shade. Widespread in the British Isles and doing well, but absent from much of central Scotland. Occurs throughout northern mainland Europe as far north as the southern half of Scandinavia. In southern Europe the cream spots are all orange, making it look more like a Wall (see p. 154) or fritillary.

SIMILAR SPECIES Quite distinctive in its normal northern form.

Speckled Wood in its paler more orange southern Form, in France.

Pupa of Speckled Wood on grass.

Wall

Lasiommata megera

Wingspan 44–50mm. Also known as the Wall Brown, this pretty little butterfly is a familiar sight in warm coastal areas, although it has declined alarmingly in recent years. Males and females are essentially similar, with males being very slightly smaller than females on average. The sexes are broadly similar in pattern. The uppersides have a bright orange ground colour, rather like that of a fritillary, with dark brown borders, wavy brown stripes and a dark patch on the hindwing towards the body. There is a single large, dark, white-centred eye-spot on the forewing, and three or four smaller eye-spots around the rear margin of the hindwing. Males have an additional conspicuous dark brown diagonal scent brand on the forewing. The undersides are an attractive grey-brown marbled pattern, with the same eye-spot pattern as on the upper surfaces.

Wall butterfly settled on grass showing underwings.

Female Wall Brown feeding.

Male Wall Brown settled on ground, basking.

Males are more likely to be seen than females because of their habit of basking in warm, bare areas (for example along paths), and usually just moving to the next bare spot if they are disturbed. They may actively search out females if the weather is warm, or simply bask and wait for them in suitable places. Mated females lay eggs on various grasses, especially Cocksfoot, fescues, Tor Grass and Yorkshire Fog, although females are very particular about which plants they use. They avoid dense turf, and seek out grasses in broken ground, on edges or in large, dense tussocks, and the eggs may be laid on leaves, stems or exposed roots.

FLIGHT PERIOD Normally flies in two generations, late April–late June, then again July–September. In warm years (or further south) there may be a small late-autumn emergence.

HABITAT AND DISTRIBUTION Unlike the Speckled Wood (see p. 150), this is a strongly sun-loving butterfly, occurring in sunny grasslands and slopes, old quarries, railway embankments, and even wasteground and roadsides, as long as there is some broken ground and bare patches. It is widespread in Britain and Ireland as far north as southern and western Scotland, although many of the inland populations are declining or extinct, and it is becoming a largely coastal species. In northern mainland Europe it is widespread and locally common as far north as southern Scandinavia, and throughout southern Europe.

Wall butterfly egg, on grass.

SIMILAR SPECIES Rather similar in appearance to a small fritillary, although the eye-spots are a helpful distinction. In southern Europe the southern forms of the Speckled Wood are very similar, but the scalloped edges of the latter and its tendency to be in semi-shaded areas help distinguish it.

Larva of Wall, on grass.

Large Wall Brown

Lasiommata maera

Wingspan to 56mm. As its name suggests, this species is rather larger than the Wall (see p. 154), although in general it does not look especially like it. The uppersides are moderately dark brown, with a large, dark, white-centred eye-spot in an orange patch on the forewing; the extent of the orange varies from just a patch surrounding the spot, to an extensive streak curving around the wing, and northerly populations generally have the least orange. On the hindwings there are three smaller eye-spots in orange patches. The underside is very similar to that of the Wall. Females lay eggs on a variety of grasses.

FLIGHT PERIOD Flies in a single brood in much of northern Europe, June–August, although further south there are two broods.

HABITAT AND DISTRIBUTION Occurs in a wide variety of sunny, drier grassy areas with flowers, including meadows, rocky hillsides and woodland clearings. Widespread almost throughout Europe except the far north, the Low Countries and the British Isles.

SIMILAR SPECIES The Northern Wall Brown *L. petropolitana* is extremely similar. It is slightly smaller and darker, and there is a wavy transverse line on the hindwing, midway between the margin and the body – sometimes denoting a change of colour – that is not present in Large Wall. Occurs in southern Scandinavia and the Baltic states, and further south in mountain areas.

Large Wall Brown basking on rock.

Gatekeeper

Pyronia tithonus

Wingspan 38mm (smallest males), to 48mm (largest females). A pretty little butterfly, also known as the Hedge Brown. The sexes are similar in general pattern, although females are slightly larger than males. The upper surface is predominantly orange, edged with a broad, grey-brown band, and with an extensive patch of grey-brown next to the body on the hindwing. There is a large, round brown spot on the forewing tip, with two little white eye-spots, and a single smaller, similar spot on the hindwing with one white eye-spot. In females most of the forewing is clear orange apart from the border and spot, but males have a conspicuous broad dark sex brand spreading transversely part or all the way across the middle of the orange patch. Underneath they are patchily brown and cream with a few small, white-centred eye-spots. Populations in southern Europe are much paler underneath than northern populations.

Gatekeeper settled, wings partly closed, showing undersides.

Male Gatekeeper, basking on bracken.

Female Gatekeeper, basking.

Gatekeepers spend a good deal of time basking and feeding in their favoured habitats. Although they do feed on honeydew in trees, they mainly visit flowers. Because of their unusually short tongues (about 6mm long) they feed almost entirely on open, flat flowers that yield their nectar readily, such as Bramble, Ragwort and thistles. The eggs are laid on various fine grasses, including bents, fescues, meadow-grasses and others, usually growing in sunny, sheltered positions by shrubs.

FLIGHT PERIOD Invariably single brooded, even in southerly areas, flying late June–early September.

Larva of Gatekeeper, on grass.

Gatekeeper settled with wings fully closed, showing hindwings only.

HABITAT AND DISTRIBUTION Although this species occurs in quite a wide variety of habitats, these are all defined by being warm, sheltered, grassy, flowery and with shrubs. Common in scrub, woodland rides, edges and clearings, or even along old hedgerows with a flowery verge; now more frequent as a garden butterfly than it used to be. In Britain it is often abundant where it does occur – in most of England and Wales, and southernmost Ireland – but it is rare or absent elsewhere. At present it seems to be doing well, and is slightly extending its range northwards, perhaps in response to milder winters. In northern mainland Europe it extends no further north than north-west Germany, and is absent from Scandinavia and the Baltic states. It is probably too sensitive to cold winters.

SIMILAR SPECIES Most often confused with the Meadow Brown (see p. 166), but the Gatekeeper is smaller, with more orange, two white spots in the forewing eye-spot (usually one in the Meadow Brown), and the Meadow Brown has no white spots on the underside.

Ringlet

Aphantopus hyperantus

Wingspan 45–52mm. A medium-sized, dark and undistinguished, but rather pretty butterfly. The sexes are quite similar, although females are slightly larger, as is normal for this group. The upper surfaces are rich chocolate-brown in newly emerged males, rather paler in females. The fringes are bright white, and there are three dark eye-spots on each forewing, and two on each hindwing. These are more conspicuous on females than males, mainly because the ground colour surrounding them is paler. Northern Scotland forms are smaller and paler, with reduced rings. Ringlets are quite active butterflies, even in dull weather, although they are not strong fliers or particularly mobile. They regularly visit flowers for nectar, especially Brambles, thistles and knapweeds. The food plants consist of various grasses, including Couch Grass and Tufted Hair Grass, although the life cycle is not fully known.

FLIGHT PERIOD Flies in a single generation, early or mid-June–early August.

HABITAT AND DISTRIBUTION Quite a shade-tolerant butterfly, avoiding the warmest, driest places, and most likely to be found in woodland glades, margins, rides and scrub, and along

Female Ringlet feeding on bramble flowers.

Male Ringlet basking on bracken.

Female Ringlet basking.

hedgerows, especially on heavier soils. Common and widespread in most of Britain and Ireland, apart from north-west England and northern Scotland. In northern mainland Europe, common as far north as central Scandinavia and Finland.

A striking form of Ringlet, known as *lanceolata*.

Meadow Brown

Maniola jurtina

Wingspan 44–56mm. Although hardly the most beautiful or popular of our butterflies, the Meadow Brown is certainly one of the most common and best known. Females are larger than males, and the sexes are distinctly different from each other. On the uppersides males are largely dark chocolate-brown, with a small, white-centred dark eye-spot – which has just one white spot – in a patch of orange on the forewing. Females have the same eye-spot, but a much larger area of orange on the forewing, and some orange on the hindwing, making them look much more orange overall. Underneath, males and females are similar, with an orange forewing with an eye-spot, and a duller brown and cream hindwing, with one or more dark spots towards the margin. Unlike many butterflies, Meadow Browns can fly in quite dull, cool weather; they spend much of their time visiting scabious, knapweed and thistle flowers for nectar. The eggs are laid on various grasses, especially meadow-grasses and fescues.

The relationship between the genetics, appearance and behaviour in this species has been studied in some detail. The

Two male Meadow Browns feeding on Spear Thistle.

Female Meadow Brown basking.

Meadow Brown male feeding on Fleabane.

Underside of male Meadow Brown.

markings of Meadow Browns vary considerably between sexes and individuals. It has been shown that various characteristics are linked under the control of groups of genes, and that the black spots on the under hindwings are linked to other characteristics that affect the butterflies' survival rates. Males are believed to have a duller colour but more spots because they spend more time on the wing than females; this colouration gives them better camouflage during flight. Females, by contrast, need to be well camouflaged at rest, and additionally have flash colours that may frighten a potential predator when it is close. Similarly, populations living in different habitats or climates may do better with a different wing pattern.

FLIGHT PERIOD Flies in a single generation but with protracted emergence, in late May–September. In southern mainland Europe individuals may aestivate (pass the summer in a torpid state), but do not do so in the north.

Underside of female Meadow Brown feeding on Hemp Agrimony.

HABITAT AND DISTRIBUTION Frequently abundant in warm, rough grasslands such as ungrazed downland, rough meadows, wasteground, woodland clearings, stable dunes and roadsides. Occasionally, huge numbers may be found under ideal conditions. Widespread throughout the UK and Ireland except for the highest mountains. Common in northern mainland Europe, but only as far north as southern Scandinavia and Estonia. Currently slowly expanding its range northwards.

SIMILAR SPECIES Females are often confused with Gatekeepers, although Meadow Browns are much larger (for differences, see p. 163). Males might be confused with Ringlets (see p. 164), but these latter are smaller and darker, and have more eye-spots on the wings and a whiter fringe.

Small Heath

Coenonympha pamphilus

Wingspan 29–35mm. Despite being widespread and common, this is one of our most inconspicuous and least-noticed butterflies, especially as it always settles with its wings closed. Males and females are generally similar, with males being smaller on average. The species is not very different in size from a hairstreak or skipper. The male's uppersides are dull orange-brown with an inconspicuous eye-spot on the forewing; the female is paler, although it is almost impossible to see the uppersides. The undersides are orange (male) or yellowish-brown (female), tipped with grey-brown and with a single small, neat eye-spot towards the tip. The hindwing is greyish-brown, banded with darker and lighter shades. When concealment is important the forewings are hidden; the orange forewing with its confusing eye-spot may then be flashed if a predator comes close.

Small Heath settled on dry grass.

Mating pair of Small Heaths on Fragrant Orchid.

Small Heaths have a low, floppy flight, unlike that of most other species, and settle readily on the ground. Males gather communally in a lek, usually in grassland with bare ground close to a bush, in order to have the best chance of attracting females. The males constantly interact and jostle to get the best perches, and those that do are most likely to mate. After mating the females move away from the males, seeking out the food plants, which are various grasses such as meadow-grasses, bents and fescues.

FLIGHT PERIOD Likely to be seen on the wing late April–early October in warmer areas, in a series of partly overlapping generations. Further north, for example in Scotland, the flight period is shorter, with often just one brood, June–August.

Underside of roosting Small Heath.

HABITAT AND DISTRIBUTION A very widespread species in the UK, occurring almost everywhere, including in areas of rough grassland in mountains. Particularly favoured areas include chalk downland, sand dunes and other coastal grasslands, grassy heaths, old quarries and clearings in woodland. The species has declined in numbers recently, although it is still widespread. It is common throughout most of northern mainland Europe, except northern Scandinavia.

SIMILAR SPECIES The Large Heath and Pearly Heath (see pp. 174 and 173) are both rather similar.

Pearly Heath

Coenonympha arcania

Wingspan 34–40mm. This species is similar in shape and style to the Small Heath (see p. 170), but is distinctly larger and has more boldly marked wings. From above, the forewings are deep orange strongly edged with dark brown, while the hindwings are dark brown. Underneath – the side that is normally seen – it is much more conspicuously marked. The forewing underside resembles the upperside, but the hindwing is reddish-brown with a bold, variably wide white stripe across the wing; on the margin side of this there are three or four eye-spots, with (often) one eye-spot in the white band, and another just on the body side of it. Within northern continental Europe this combination is readily recognisable; further south there are more similar species.

FLIGHT PERIOD Usually single brooded, late May–August.

HABITAT AND DISTRIBUTION
Grassy places such as damp meadows and woodland clearings. Absent from Britain. Widespread in southern and eastern mainland Europe, reaching north-west to Belgium and southern Sweden.

SIMILAR SPECIES
Other heaths.

Pearly Heath settled, showing the distinctive white band on hindwing.

Large Heath

Coenonympha tullia

Wingspan 40–44mm. Not surprisingly, this is like a larger version of the Small Heath (see p. 170). It is highly variable in markings over its geographical range, and many local races have been described. Generally, the upper surface is pale orange-brown with a variable number of pale eye-spots on both wings, although – like other heaths – this species almost invariably settles with its wings closed. Underneath, the paler and least-marked forms – typically from Scotland and Scandinavia – look very like Small Heaths. Elsewhere there may be as many as six conspicuous eye-spots in a submarginal arc on the hindwing, with a variable but noticeable blotch of creamy-white; the forewing may have two or three spots, and a small patch of white.

Recent research has shown how the wing patterning variation relates to different lifestyles. The southerly forms tend to fly much more due to warm weather, and bold eye-spotting is a good survival aid as birds tend to attack the eye-spot and miss the body. More sedentary northern populations do well by remaining wholly camouflaged on vegetation. The food plants are White-beaked Sedge and Cotton-grass.

FLIGHT PERIOD Usually single brooded, mid-June–mid-August.

Large Heath, form *polydama*, the predominant form in Ireland and north England.

Large Heath, form *davus*, the predominant form in the rest of England.

HABITAT AND DISTRIBUTION Quite definitely a species of wet, acidic, boggy areas, such as blanket bog and damp moorland, up to a considerable altitude, where the food plants grow. Entirely northern and western in Britain, from Shropshire northwards; local but widespread in Ireland. In mainland Europe quite common from Germany eastwards and in eastern Scandinavia.

SIMILAR SPECIES Other heaths.

Large Heath, ssp. *scoticus*, the predominant form in Scotland.

Scotch Argus

Erebia aethiops

Wingspan 40mm (males), to 50mm (females); size variable according to location. A very attractive butterfly, deep chocolate-brown in colour, enlivened with bold eye-spots and orange patches. The sexes are slightly different from each other. Males are very dark brown on the upperside, with a broad, continuous band of orange on the forewings containing three eye-spots, and a discontinuous band on the hindwings containing four eye-spots. Females are lighter brown with, normally, four eye-spots on the forewings, and weakly chequered wing fringes. Underneath there are three eye-spots on the forewing in a patch of orange, while the hindwing is duller with stripes of shades of brown and grey, and paler in females than males. Males fly readily over grassland looking for females, and both sexes regularly visit flowers. The food plants are Purple Moor-grass and Blue Moor-grass.

FLIGHT PERIOD Single brooded, flying late July–early September.

Male Scotch Argus settled in wet weather.

HABITAT AND DISTRIBUTION Mainly damp upland or northern grassland in sheltered valleys, around woodland or along coasts, and up to an altitude of at least 500m. Within Britain locally common in Scotland, declining in the south and with a few sites in north-west England, but absent from the rest of England, Wales and Ireland. In mainland Europe essentially a mountain species, absent from Scandinavia, but more common from central Germany southwards.

Male Scotch Argus basking.

Female Scotch Argus, basking.

SIMILAR SPECIES Most likely to be confused with the Mountain Ringlet (for distinctions, see p. 178).

Mountain Ringlet

Erebia epiphron

Wingspan 35–40mm. A rather inconspicuous little butterfly, generally brown in colour, and made more inconspicuous by its habit of only flying on the sunniest days and by living in remote mountain areas. Males are slightly smaller than females, and both sexes are generally brown above, with lines of black spots within a dull orange band on both the hindwing and forewing. Females are paler in overall colour, and have larger, brighter orange spots. Rather surprisingly for a butterfly that lives in the least sunny areas, it is only active in sunny weather and almost completely disappears on dull days. Females fly very little at all. The food plant in the UK is almost exclusively Mat-grass, although elsewhere other moorland grasses are used.

FLIGHT PERIOD Mid-June–early August in one generation, although variable according to site, and generally becoming earlier.

HABITAT AND DISTRIBUTION Occurs in usually isolated colonies on moorland and upland grassland; most common where there is some shelter. Uncommon in the UK and restricted to the Lake District and parts of the Scottish Highlands. Absent from much of the rest of northern Europe, but quite common on mountains from the Vosges southwards.

Mountain Ringlet settled to roost at dusk.

SIMILAR SPECIES Rather similar to the Scotch Argus (see p. 176), but smaller and duller, and the wing-spots are solid black without a white eye-spot in each. In alpine areas of Europe there are many rather similar ringlets that need close examination for identification.

Male Mountain Ringlet in Lake District.

Female Mountain Ringlet among Thyme flowers.

Marbled White

Melanargia galathea

Wingspan 53–58mm. A conspicuous, distinctive and popular butterfly, redolent of summer grasslands and warm days. Despite its name and appearance, it is more closely related to the browns than the whites. Males and females are broadly similar in appearance, with subtle differences. Females are generally slightly larger than males. The upper surfaces are boldly marked with black-and-white chequering, with the two colours covering roughly the same proportion of the surface, and there are no obvious eye-spots. Underneath they are much paler and more muted, with an arc of spots around the hindwing margin, each with a pale blue centre. Although there is wide variation in colour, in general males are more greyish underneath than females, and females are greenish-brown. Marbled Whites are very active butterflies, with both sexes flying regularly in warm weather, especially the males, which fly almost constantly in search of females. In cool sunshine they bask readily, and in very hot midday sunshine they settle with closed wings to prevent overheating. Both sexes readily visit flowers such as knapweeds and scabiouses for nectar. The food plants include various grasses, but especially Red Fescue; the female seeks this out, then casually deposits her eggs on and around the plant.

A noticeable feature of many Marbled Whites is the presence of small red mites attached to the body. These are the parasitic larval stages of *Trombidium breei*, which feed on the blood of the

Pupa of Marbled White.

Female Marbled White on Creeping Thistle.

Male Marbled White on Welted Thistle.

Mating pair of Marbled Whites, with female on the right.

butterfly; in small numbers they do little harm, but heavy infestations can be damaging or may even kill the butterfly. Males tend to be more affected than females. Other butterflies suffer, too, although the mites are most conspicuous on Marbled Whites.

FLIGHT PERIOD Flies in a single, rather short generation, late June–early August – a high-summer butterfly par excellence.

HABITAT AND DISTRIBUTION A butterfly of rough, flowery, dry grassland, such as chalk and limestone downland, although it also occurs on heavier soils in meadows and pastures. It is found in defined colonies and is not particularly mobile, but it does colonise new sites under good conditions. Within Britain locally common in southern and eastern England, very local in South Wales and absent elsewhere, although it is currently expanding its range. In mainland Europe common from northern Germany and Poland southwards.

SIMILAR SPECIES Can be distinguished from the whites by the much greater amount of black, and the chequered appearance. Further south there are several similar close relatives, although none occurs in northern Europe.

Grayling

Hipparchia semele

Wingspan 52–62mm. A large but inconspicuous butterfly – the largest of the browns in the UK. Males and females are similar (females are slightly larger). The uppersides (seen only in flight) are brown, with extensive areas of pale brown, cream and orange containing a few black eye-spots. On the undersides the forewings are pale orange edged with dark grey, with two dark, white-centred eye-spots; the hindwings are fuzzily marbled with shades of grey and brown. At rest the Grayling always closes its wings, and only shows the forewings briefly or when threatened, so it is well camouflaged. It also constantly adjusts the angle of the wings to the sun, which achieves both temperature regulation and camouflage. The food plants are fine grasses, especially bents and fescues.

FLIGHT PERIOD Flies only in a single, rather late generation, early July–early September, although there is a smaller variant found in north-west England and Wales that flies about a month earlier.

Grayling butterfly settled on the ground.

Grayling feeding on Marjoram.

HABITAT AND DISTRIBUTION May be found on a variety of dry, grassy sites. Once widespread throughout Britain, but now virtually confined to lowland heaths and coastal habitats, with populations all around the coast of Britain; rare inland. In northern mainland Europe locally common from southern Scandinavia and Estonia southwards.

SIMILAR SPECIES Distinctive in Britain and most of northern Europe. In France the False Grayling *Arethusana arethusa* is very similar, but often settles with wings open, and has only one forewing eye-spot underneath and a more distinct white stripe on the hindwing. Rare, from northern France southwards.

Grayling settled on the ground, demonstrating its cryptic colouring.

Mating pair of Graylings on gorse bush.

FOOD PLANTS

Common Bird's-foot Trefoil *Lotus uliginosus*
Horseshoe Vetch *Hippocrepis comosa*
Wild Strawberry *Fragaria vesca*
Cinquefoils *Potentilla* spp.
Purple Moor-grass *Molinia caerulea*
Wood False-brome *Brachypodium sylvaticum*
Hollyhocks *Alcea* spp.
Hairy Violet *Viola hirta*
Timothy *Phleum pratense*
Yorkshire Fog *Holcus lanatus*
Cocksfoot *Dactylis glomerata*
Creeping Soft-grass *Holcus mollis*
Marjoram *Origanum vulgare*
Greater Knapweed *Centaurea scabiosa*
Tor Grass *Brachypodium pinnatum*
Sheep's Fescue *Festuca ovina*
Milk Parsley *Thyselium palustre*
Ragged Robin *Lychnis flos-cuculi*
Red Campion *Silene dioica*
Buddleia *Buddleia davidii*
Fennel *Foeniculum vulgare*
Wild Carrot *Daucus carota*
Wild Angelica *Angelica sylvestris*
Blackthorn *Prunus spinosa*
Hawthorn *Crataegus monogyna*
Meadow Vetchling *Lathyrus pratensis*
Bitter Vetch *Lathyrus linifolius*
Cuckoo Flower or Lady's Smock *Cardamine pratensis*
Garlic Mustard *Alliaria petiolata*
Water-cress *Nasturtium officinale*
Sea Radish *Raphanus raphanistrum* ssp. *maritimus*
White Mustard *Sinapis alba*
Hedge Mustard *Sisymbrium officinale*
Wild Mignonette *Reseda lutea*
Lucerne *Medicago sativa*
Buckthorn *Rhamnus cathartica*
Alder Buckthorn *Frangula alnus*
Cowslip *Primula veris*
Primrose *Primula vulgaris*
Sorrel *Rumex acetosa*
Sheep's Sorrel *Rumex acetosella*
Great Water Dock *Rumex hydrolapathum*
Ash *Fraxinus excelsior*
Bramble, Blackberry *Rubus* spp.
Wild Parsnip *Pastinaca sativa*
Wayfaring Tree *Viburnum lantana*
Gorse *Ulex europaeus*
Bilberry *Vaccinium myrtillus*
Wild Thyme *Thymus serpyilum*
Ragwort *Jacobaea vulgaris*
Wild Privet *Ligustrum vulgare*
Common Rock-rose *Helianthemum nummularium*
Kidney Vetch *Anthyllis vulneraria*
Holly *Ilex aquifolium*
Ivy *Hedera helix*
White Melilot *Melilotus albus*
Red Clover *Trifolium pratense*
Marsh Gentian *Gentiana pneumonanthe*
Cranesbills *Geranium* spp.
Storksbills *Erodium* spp.
Crown Vetch *Securigera varia*
Common Dog-violet *Viola riviniana*
Field Pansy *Viola arvensis*
Wild Pansy *Viola tricolor*
Marsh Violet *Viola palustris*
Knapweeds *Centaurea* spp.
Ribwort Plantain *Plantago lanceolata*
Hoary Plantain *Plantago media*
Buckshorn Plantain *Plantago coronopus*
Cow-wheat *Melampyrum pratense*
Speedwells *Veronica* spp.
Foxglove *Digitalis purpurea*
Stinging Nettle *Urtica dioica*
Thistles *Cirsium* spp. and *Carduus* spp.
Hop *Humulus lupulus*
Wall Pellitory *Parietaria judaica*
Annual Nettle *Urtica urens*
Stonecrop *Sedum spectabile*
Wild Cherry *Prunus avium*
Aspen *Populus tremula*
Blue Moor-grass *Sesleria caerulea*
Common Honeysuckle *Lonicera periclymenum*
Fly Honeysuckle *Lonicera xylosteum*
Snowberry *Symphoricarpus albus*
Goat Willow *Salix caprea*
Tor Grass *Brachypodium pinnatum*
Couch Grass *Elytrigia repens*
Tufted Hair Grass *Deschampsia cespitosa*
Bents *Agrostis* spp.
White-beaked Sedge *Rhynchospora alba*
Cotton-grass *Eriophorum* spp.
Mat-grass *Nardus stricta*
Red Fescue *Festuca rubra*

GLOSSARY

antennae (singular antenna) paired, often long sensory appendages on the head of an insect, used for feeling and smelling.
brood see generation.
caterpillar see larva.
endemic restricted in distribution to a specific area, often small.
food plant/s the plant (or plants) on which a given butterfly species lays its eggs for the caterpillars to feed on. See also nectar plant.
forewings front pair of wings.
generation in temperate areas, butterflies tend to emerge in defined partially synchronised generations rather than continuously; normally, the offspring of the first generation form the next generation.
hindwings see underwings.
host plant see food plant.
larva (plural larvae) non-flying stage of the life cycle of butterflies, which concentrates particularly on feeding; also called caterpillar.
lek process in which males gather in one more or less defined site to attract and compete for females.
metamorphosis all insects undergo big changes in their lives; those with complete metamorphosis have the stages egg – larva – pupa – adult.
mud puddling process in which butterflies, normally males, gather together to suck up liquids to provide them with substances necessary for their growth and reproduction. In good sites in hot weather the numbers can be spectacular.
nectar plants the plants that are regularly visited by butterflies for nectar; not all flowers are suitable for this purpose.
scales minute overlapping 'tiles' of tissue that give butterfly wings their colour.
scent brands see sex brands.
sex brands patches of darker scales on wings of male butterflies that release substances attractive to females; also called scent brands.
undersides underneath of wings, visible when a butterfly closes its wings.
underwings rear pair of wings; also called hindwings.

BIBLIOGRAPHY

British Wildlife magazine, www.britishwildlife.com/themagazine.asp. An excellent magazine on all aspects of British nature.

Chinery, M. *Insects of Britain and Western Europe*, 3rd edn. Bloomsbury, 2012.

Chinery, M. *Collins Complete Guide to British Insects*. Collins, London, 2005.

Gibbons, Bob. *Field Guide to the Insects of Britain and Northern Europe*. Crowood Press, Wiltshire, 1995.

Sterling, Phil et al. *Field Guide to the Micro-moths of Great Britain and Ireland*. British Wildlife Publishing, Dorset, 2012.

Tampion, John & Maureen. *Attracting Butterflies to Your Garden*. Guild of Master Craftsmen Publications, Lewes, 2011.

Thomas, Jeremy. *The Butterflies of Britain and Ireland*, 2nd edn. British Wildlife Publishing, Dorset, 2010.

Waring, Paul et al. *Field Guide to the Moths of Great Britain and Ireland*. British Wildlife Publishing, Dorset, 2003.

RESOURCES AND ORGANISATIONS

Useful Websites

UK Butterflies
www.ukbutterflies.co.uk/index.php for species information, sites, news and current activities.

British Butterflies
www.britishbutterflies.co.uk is similar.

For European species
Matt's European Butterflies, www.eurobutterflies.com, and Guy Padfield's European Butterflies, www.guypadfield.com, are both useful, as is Roger Gibbons's (no relation!) Butterflies of France, www.butterfliesoffrance.com.

Organisations

Buglife
www.buglife.org.uk
Join to help conserve insects, or visit the website for species identification, surveys and campaigns.

Butterfly Conservation
www.butterfly-conservation.org
Join to help conserve and learn about butterflies and moths. Lots of local events. The primary organisation in the UK for butterfly conservation; also with European links and a European special-interest group. Runs regular conferences and events.

Field Studies Council
www.field-studies-council.org
Courses on all aspects of nature and the environment, including insects.

The Kingcombe Centre, Dorset
www.kingcombe.org
Various courses or accommodation in the middle of a lovely insect-rich area.

The Wildlife Trusts
www.wildlifetrusts.org
The umbrella organisation for all the county Wildlife Trusts. They play a crucial role in conserving habitats for insects and all wildlife throughout the UK.

ACKNOWLEDGEMENTS

The author would like to express particular thanks to Peter Marren, for reading the text and offering many helpful suggestions, though any remaining errors are entirely my fault. And to Richard Revels for his endless patience in searching through his huge collection of excellent butterfly photographs to find the right images.

IMAGE CREDITS

Bloomsbury Publishing would like to thank the following for providing photographs and for permission to produce copyright material. While every effort has been made to trace and acknowledge all copyright holders, we would like to apologise for any errors or omissions and invite readers to inform us so that corrections can be made in any future editions of this book.

Key = t = top, m = middle, b= bottom

Bob Gibbons: 10, 11, 14, 16, 20, 21b, 25, 27b, 31, 35, 36, 37t, 37b, 38b, 39t, 43t, 45, 49, 52, 53, 55t, 57t, 57b, 61b, 62, 65b, 66, 69b, 73, 75b, 76, 77, 80, 87, 88, 90, 91t, 91b, 94, 95b, 98, 99t, 107, 110, 111b, 112, 117, 118, 121t, 122, 123t, 123b, 125, 126, 127t, 127b, 131b, 135b, 139t, 139b, 143, 147, 148t, 148b, 149b, 152, 153t, 154, 159, 160, 161t, 163, 164, 165t, 166, 167t, 168, 169, 170, 172, 176, 177t, 178b, 181, 182, 185t, 185b

Richard Revels: 12, 13, 15, 17, 18, 19, 21t, 22, 23, 24, 26, 27t, 28, 29, 30, 32, 33, 34, 38t, 39b, 40, 41, 42, 43b, 44, 46, 47t, 47b, 48, 51t, 51b, 54, 55b, 56, 58, 59t, 59b, 60, 61t, 61m, 63t, 63b, 64, 65t, 67, 68, 69t, 70, 71, 72, 74, 75t, 78, 79, 81, 82, 83, 84, 85, 86, 89t, 89b, 92, 93t, 93b, 95t, 96, 97t, 97b, 99b, 100, 102, 103, 104, 105t, 105b, 106t, 106b, 108, 109t, 109b, 111t, 113t, 113b, 114, 115, 116, 119, 120, 121b, 124, 129t, 129b, 130, 131t, 133t, 133b, 134, 135t, 136, 137, 138, 140, 141t, 141b, 142, 144t, 144b, 145t, 145b, 146t, 146b, 149t, 150, 153b, 151, 155t, 155b, 156, 157, 161, 162, 165m, 165b, 167b, 171, 173, 174, 175t, 175b, 177b, 179t, 179b, 180, 181t, 183, 184

Shutterstock: 101

INDEX

INDEX

INDEX